Six Themes on Variation

STUDENT MATHEMATICAL LIBRARY
Volume 26

Six Themes on Variation

Robert Hardt, Editor

Steven J. Cox
Robin Forman
Frank Jones
Barbara Lee Keyfitz
Frank Morgan
Michael Wolf

AMERICAN MATHEMATICAL SOCIETY

Editorial Board

Davide P. Cervone Robin Forman
Daniel L. Goroff Brad Osgood
Carl Pomerance (Chair)

Cover art provided by Matthias Weber. Used with permission.

2000 *Mathematics Subject Classification.* Primary 49–01.

For additional information and updates on this book, visit
www.ams.org/bookpages/stml-26

Library of Congress Cataloging-in-Publication Data
Six themes on variation / Steven J. Cox... [et al.] ; Robert M. Hardt, editor.
 p. cm. — (Student mathematical library v. 26)
 Includes bibliographical references.
 ISBN 0-8218-3720-6 (alk. paper)
 1. Calculus of variations. I. Cox, Steven J. (Steven James), 1960– II. Hardt, R. (Robert), 1945– III. Series.

QA315.S59 2004
515'.64—dc22   2004059479

Copying and reprinting. Individual readers of this publication, and nonprofit libraries acting for them, are permitted to make fair use of the material, such as to copy a chapter for use in teaching or research. Permission is granted to quote brief passages from this publication in reviews, provided the customary acknowledgment of the source is given.

Republication, systematic copying, or multiple reproduction of any material in this publication is permitted only under license from the American Mathematical Society. Requests for such permission should be addressed to the Acquisitions Department, American Mathematical Society, 201 Charles Street, Providence, Rhode Island 02904-2294, USA. Requests can also be made by e-mail to reprint-permission@ams.org.

© 2004 Robert Hardt. All rights reserved.
Chapters copyrighted individually by the authors.
Printed in the United States of America.

∞ The paper used in this book is acid-free and falls within the guidelines established to ensure permanence and durability.
 Visit the AMS home page at http://www.ams.org/
  10 9 8 7 6 5 4 3 2 1   09 08 07 06 05 04

Contents

Preface vii

List of Contributors xi

Calculus of Variations: What Does "Variations" Mean?
    Frank Jones 1

How Many Equilibria Are There? An Introduction to Morse Theory
    Robin Forman 13

Aye, There's the Rub. An Inquiry into Why a Plucked String Comes to Rest
    Steven J. Cox 37

Proof of the Double Bubble Conjecture
    Frank Morgan 59

Minimal Surfaces, Flat Cone Spheres and Moduli Spaces of Staircases
    Michael Wolf 79

Hold That Light! Modeling of Traffic Flow by Differential Equations
    Barbara Lee Keyfitz 127

Preface

One November, Rice University hosted a group of thirty undergraduate mathematics majors with the purpose of introducing them to research mathematics and graduate school. The principle part of this introduction was the series of talks and workshops, which all took up some idea or theme from the calculus of variations. These were so successful that the American Mathematical Society encouraged us to present them to a wider audience, in the form you see here.

The calculus of variations is a beautiful subject with a rich history and with origins in the minimization problems of calculus (see Chapter 1). Although, as we will discover in the chapters below, it is now at the core of many modern mathematical fields, it does not have a well-defined place in most undergraduate mathematics courses or curricula. We hope that this small volume will nevertheless give the undergraduate reader a sense of its great character and importance.

An interesting story motivating the calculus of variations comes from Carthage in 900 BC, long before the discovery of calculus by Newton and Leibniz. Queen Dido, as a result of a bargaining negotiation, obtained "as much land as could be enclosed by the skin of an ox." She had the ox skin cut into strips as thin as practically possible and formed a long cord of fixed length. If her choice of land had been restricted to flat inland territory, then she would presumably have chosen a large circular region. This is because the circle, among all planar closed curves of fixed length, encloses the maximum area. But she had the choice of territory with a flat coastline and cleverly chose a semi-circular region, with the cord's endpoints on the shoreline. This gives more area and is actually the mathematically optimal solution. A change or *variation* of the shape of the cord cannot give a new region of greater area.

Calculus of Variations arises when one *differentiates*, in the sense of the calculus of Newton and Leibniz, a one-parameter family of such variations. This first occurs in the works by P.L.M. de Maupertuis (1698-1759), G.W. Leibniz (1646-1716), Jakob Bernoulli (1654-1705), Johann Bernoulli (1667-1748), L. Euler (1707-1783), and J.L. Lagrange (1736-1813). It has historically largely been the study of optimal paths, for example as a geodesic curve in a space or as a path of least action in space-time. See the nice presentation in Chapter 1 of *The Parsimonious Universe: Shape and Form in the Natural World* by S. Hildebrandt and A. Tromba (Copernicus, New York, 1996).

In modern language, the birth of the calculus of variations occurs in the transition from the study of a critical point of a function on a line (as in calculus) to that of a critical curve or critical surface for a functional, such as length or area, on an infinite-dimensional space of such objects. As discussed in Chapter 1 by Frank Jones, the condition of criticality for these objects leads to the important partial differential equations of Euler and Lagrange. Various physical problems also give rise to natural conditions constraining the space of admissible objects. One such constraint involves a fixed boundary, as with a classical vibrating string or a soap film spanning a wire. Another constraint is seen in Queen Dido's problem. Her problem may be equivalently reformulated as the *isoperimetric problem* of finding a curve of minimum length enclosing a given fixed area. The analogous two-dimensional isoperimetric problem of finding a surface of least area enclosing a given volume (or volumes) occurs in soap bubble models.

In Chapter 2 by Robin Forman, one considers the connection between such critical or equilibrium points and the topology or geometry of the ambient spaces. Here is a quick elegant introduction to the simple, but subtle, ideas of Marston Morse from the 1940s.

Preface

Their generalizations involving infinite-dimensional spaces of paths (or solutions of other PDEs) have had a profound influence on 20th century mathematics. One here encounters critical paths that may not be globally or even locally length minimizing. For example, the "ridge trail" over a mountain range is a length-critical path that is *unstable*. Some slight variation may give a (more dangerous) path of shorter length.

Physicists and mathematicians have long been interested in understanding and modeling vibrating strings, as in bowed or plucked instruments. Steve Cox discusses in Chapter 3 the cause of the observed decay of the amplitude. Such decay is usually neglected in introductory treatments in physics courses. His chapter well illustrates the full range and difficulty of scientific inquiry from acquiring experimental data, to synthesizing data, to mathematical modeling, to finding actual or approximate solutions. The discussion here includes a useful introduction and illustration of the classical "Principle of Stationary Action".

The isoperimetric problem that a surface of least area in space enclosing a single given volume must be an ordinary round sphere was solved rigorously over 100 years ago. It was claimed by Archimedes and Zenodorus in antiquity, but proved by H. Schwarz in 1884. At the Rice undergraduate conference, Frank Morgan discussed the important *Double Bubble Conjecture* that a surface of least area enclosing two fixed volumes consists simply of two adjoined spherical caps joined by a third spherical interface (with radii determined by the given volumes). In 1998, this conjecture was proven by M. Hutchings, F. Morgan, M. Ritoré, and A. Ros. For Chapter 4 of the present volume, Frank Morgan's original talk has been replaced by a reprint of his excellent 2001 MAA article exposing this result.

Minimal surfaces occur in the calculus of variations as critical points of the area functional and provide models for some soap films. K. Weierstrass (1815-1897) showed that they also enjoy a mathematical representation in terms of complex-valued functions. Chapter 5 by Mike Wolf explains carefully this connection and gives a related, recently discovered representation that allows the construction of several rich new families of minimal surfaces. See the many beautiful illustrations here. This chapter is a great introduction to some of the many important relationships among the calculus of variations, complex analysis, and differential geometry.

Differential equations for modeling traffic flow are derived and analyzed in the chapter by Barbara Keyfitz. The continuum model derived here is natural, consistent, and leads both to many observed familiar *discontinuous* phenomena such as shock waves and to many important open mathematical problems. It is a great example of the fruitful interplay between pure and applied mathematics. Proper careful modeling not only gives better scientific applications but reveals beautiful often hidden mathematical structures.

On Saturday afternoon of the Rice conference, students also had the opportunity to actively participate in Steve Cox's experiments with vibrating strings (see the illustrations in Chapter 3) or with Frank Morgan's soap films and soap bubbles or to hear from Robin Forman about many recent open problems in mathematics. The format of the Calculus of Variations Conference worked well, and three other similarly structured undergraduate conferences have since been held at Rice: *Low Dimensional Geometry and Topology*, *Geometric Aspects of Combinatorics*, and *Mathematical Problems in Biology*.

The editor appreciates the great patience and help of Ed Dunne of the American Mathematical Society in assembling this book and the suggestion of Carl Pomerance for the title "Six Themes on Variation".

List of Contributors

Steve Cox, Rice University
Robin Forman, Rice University
Robert Hardt, Rice University
Frank Jones, Rice University
Barbara Lee Keyfitz, University of Houston and The Fields Institute
Frank Morgan, Williams College
Michael Wolf, Rice University

Calculus of Variations: What Does "Variations" Mean?

Frank Jones

The purpose of this brief introduction to our conference is to explain some of the philosophy behind what is called "the calculus of variations". I am not going to give any proofs, and I am going to be very loose in my hypotheses. Furthermore, I am just going to give a very small taste of what sort of problems are of interest within this subject. I feel that this is justified here, as the audience is sophisticated undergraduate students, and it is rare indeed for this subject to be taught in any undergraduate course in mathematics.

The idea here is that the calculus of variations is more of a vague area of mathematical analysis than it is a well-defined subject area.

In all of the following I restrict attention to a *real-valued* function, and the sort of domain where the function is defined determines what we call our analysis. The initial idea is that we want to figure out how to locate points in the domain where the function attains a maximum or a minimum value.

©2004 by the author

I. The domain is one-dimensional

This is one of the most basic situations in all of calculus. If the function f attains a maximum or a minimum value at a point x in the interior of its domain, then of course the derivative is zero at that point:

$$f'(x) = 0.$$

As you know, this equation may be satisfied even if f does not attain an extreme value at x. A point x which satisfies this equation is called a *critical point* for the function f.

II. The domain is n-dimensional

We again suppose that the function f attains an extreme value at a point x in $\mathbb{R}^n$ and in the interior of its domain. In this case we introduce a vector h (also in $\mathbb{R}^n$) which serves as a direction for analysis of the function. Then the point $x + th$, where t is a real number, represents a point on the straight line through x in the direction h. Thus the function of t given by the expression $f(x + th)$ attains an extreme value at $t = 0$. Therefore the situation given in part I shows that

$$\frac{d}{dt}(f(x+th))\big|_{t=0} = 0.$$

The quantity on the left side of this equation is often called the *directional derivative* of f in the direction h and is denoted

$$D_h f(x).$$

The chain rule of multivariable calculus enables us to rewrite this equation in the form

(1) $$\sum_{j=1}^{n} \partial f/\partial x_j(x) h_j = 0.$$

Since (1) must hold for every choice of the direction h, we may use the n coordinate vectors $h = (0, \ldots, 0, 1, 0, \ldots, 0)$ to conclude that (1) is equivalent to

(2) $$\partial f/\partial x_j(x) = 0 \quad \text{for all } 1 \leq j \leq n.$$

Calculus of Variations

While all of this should be very familiar to you, there are some important observations that we need to stress:

- A point x satisfying (2) is said to be a *critical point* for f.
- Just as in the case $n = 1$, any extreme point for f is a critical point, but not conversely.
- The point $x+th$ used in the analysis is said to be a *variation* of the fixed point x. The idea is that $|t|$ is small, and it is only the limiting behavior of $f(x+th)$ as $t \to 0$ that interests us.
- We proceed from (1) to (2) by making judicious choices of h.

III. The domain is infinite-dimensional

Now we come to the actual situation of interest in the calculus of variations. We try to analyze the critical behavior of a real-valued function f whose domain is a certain space of real-valued functions. Thus, for each such function u there is a corresponding value $f(u)$.

We assume that f attains an extreme value at a certain function u. Then for any function φ that is "admissible" in some sense, we consider the *variation* $u + t\varphi$ of the function u. Then $f(u + t\varphi)$ has an extreme value at $t = 0$, so we conclude that the "directional derivative"

$$D_\varphi f(u) = \frac{d}{dt}(f(u + t\varphi))\,\big|_{t=0}$$

must equal zero. Thus we are led to the definition that u is a *critical "point"* for f if

$$D_\varphi f(u) = 0 \qquad \text{for all admissible functions } \varphi.$$

This is about as far as we can go without specifying just what sort of function f is. I have chosen to demonstrate the ideas with the most well-known situation.

IV. The Euler-Lagrange scenario

This is a situation that is encountered in a tremendous variety of circumstances. Without being too specific about hypotheses, suppose that D is a "reasonable" bounded open set in $\mathbb{R}^n$ with closure $\overline{D}$, and suppose g is a smooth real-valued function defined on $\overline{D} \times \mathbb{R}^{n+1}$. Then for any C^1 function $\overline{D} \xrightarrow{u} \mathbb{R}$ we can form the integral

$$f(u) = \int_D g(x, u, \partial u/\partial x_1, \ldots, \partial u/\partial x_n) dx.$$

Usually it will be the case that u has to satisfy some restrictions on the boundary ∂D of D. These restrictions are called *boundary conditions*. A common example is the so-called *Dirichlet* condition, in which the restriction of u to ∂D is a given function defined on ∂D.

We are then going to try to investigate possible critical "points" for f.

Since u must satisfy the boundary conditions, we are somewhat limited in permitted variations $u + t\varphi$ of u. These variations will certainly be allowed if the functions φ we use are infinitely differentiable on D and are zero near ∂D. We call such functions *test functions* and we write

$$\varphi \in C_0^\infty(D).$$

For each such test function we can form the directional derivative

$$\begin{aligned} D_\varphi f(u) &= \frac{d}{dt}(f(u + t\varphi))\big|_{t=0} \\ &= \frac{d}{dt} \int_D g(x, u + t\varphi, u_{x_1} + t\varphi_{x_1}, \ldots) dx \big|_{t=0} \\ &= \int_D \frac{d}{dt} g(x, u + t\varphi, u_{x_1} + t\varphi_{x_1}, \ldots) dx \big|_{t=0} \\ &\stackrel{\text{chain rule}}{=} \int_D \left(\frac{\partial g}{\partial u} \varphi + \frac{\partial g}{\partial u_{x_1}} \varphi_{x_1} + \ldots \right) dx. \end{aligned}$$

Next we integrate by parts to get rid of all the terms $\partial \varphi / \partial x_i$, noting that no integration over ∂D is required, thanks to the fact that φ is zero near ∂D. The result is

(3) $$D_\varphi f(u) = \int_D \{\partial g/\partial u - \sum_{i=1}^n \partial(\partial g/\partial u_{x_i})/\partial x_i\} \varphi \, dx.$$

In case u is a critical point for f, then by definition we conclude that

(4) $$\int_D \{-\} \varphi \, dx = 0$$

Calculus of Variations

for all test functions φ. Here of course $\{-\}$ denotes the expression in the integrand of (3). This result is the precise analog of (1) in Section II, and we would like to obtain the result similar to (2); that is, we would like to conclude that $\{-\}$ itself is equal to zero.

At this point we use a rather easy fact about integration. However, this is of such important historical significance that it actually has the rather daunting name, **The Fundamental Lemma of the Calculus of Variations**. This asserts that if (4) is valid for all test functions φ, then $\{-\} = 0$. The proof depends somewhat on the assumptions. The easiest version assumes that the function $\{-\}$ is continuous. Then if it is nonzero at some point $x_0 \in D$, say it is positive, then by continuity it remains positive in a neighborhood of x_0. Then we can select a test function φ which is ≥ 0 on D, which is positive at x_0, and which is identically zero outside a small neighborhood of x_0. Then $\{-\}\varphi$ is a continuous nonnegative function on D which is positive in a neighborhood of x_0, so that of course

$$\int_D \{-\}\varphi dx > 0.$$

This contradicts (4).

Thus we conclude that if u is a critical point for f, then

$$\boxed{\frac{\partial g}{\partial u} - \sum_{i=1}^{n} \frac{\partial}{\partial x_i}\left(\frac{\partial g}{\partial u_{x_i}}\right) = 0.}$$

This result is called the **Euler-Lagrange equation**. Notice that it is a necessary and sufficient condition for u to be a *critical point* for f. It is therefore a necessary condition for f to attain an *extreme value* at u, but nothing in the argument would even hint that it would be a sufficient condition (and it is not).

Be careful of the strange notation $\partial g / \partial u_{x_i}$. The function g depends on x, u, and n other independent real variables. Thus $\partial g / \partial u_{x_i}$ is the partial derivative of g with respect to the argument that occupies the slot where we have inserted u_{x_i}.

V. Several classical examples

In this last portion of the talk we shall present a variety of beautiful ancient examples of the use of the Euler-Lagrange equation.

A. Minimal surfaces. Given a closed curve in $\mathbb{R}^3$, the problem is to try to find a surface of minimal area which "fills in" the given curve. Let us restrict attention to the following situation. Given a domain D in the (x,y)-plane with boundary ∂D, we assume that the given curve has the form $(x, y, \gamma(x,y))$ for $(x,y) \in \partial D$, and that we seek a corresponding surface of minimal area which has the explicit description $z = u(x,y)$, where u is the unknown function. Vector calculus gives us the area of the surface in the form

$$\iint_D \sqrt{1 + u_x^2 + u_y^2}\, dx\, dy.$$

This is a perfect set-up for Euler-Lagrange, and the equation we obtain is

$$-\frac{\partial}{\partial x}\left(\frac{\partial}{\partial u_x}\sqrt{1 + u_x^2 + u_y^2}\right) - \frac{\partial}{\partial y}\left(\frac{\partial}{\partial u_y}\sqrt{1 + u_x^2 + u_y^2}\right) = 0.$$

In other words,

$$(5) \quad \frac{\partial}{\partial x}\left(\frac{u_x}{\sqrt{1 + u_x^2 + u_y^2}}\right) + \frac{\partial}{\partial y}\left(\frac{u_y}{\sqrt{1 + u_x^2 + u_y^2}}\right) = 0.$$

This equation is often called *the minimal surface equation*. Performing the indicated derivatives puts it into the form

$$(5') \quad (1 + u_y^2)u_{xx} - 2u_x u_y u_{xy} + (1 + u_x^2)u_{yy} = 0.$$

We now give three examples of solutions of this interesting partial differential equation; notice that we are effectively ignoring the initial desire to minimize surface area filling in a curve.

Example 1 (Plane). Clearly, $u(x,y) = Ax + By + C$ is a solution of the minimal surface equation, which is significant, but not all that interesting.

Calculus of Variations

Example 2 (Sherk's surface). Let us seek a solution which has the special form $u(x, y) = f(x) + g(y)$, where f and g are each functions of a single variable. Then (5′) becomes

$$(1 + g'(y)^2)f''(x) + (1 + f'(x)^2)g''(y) = 0.$$

This differential equation splits into two ordinary differential equations by separating the variables:

$$\frac{f''(x)}{1 + f'(x)^2} + \frac{g''(y)}{1 + g'(y)^2} = 0,$$

so that there must be a constant c such that

$$\frac{f''(x)}{1 + f'(x)^2} = -\frac{g''(y)}{1 + g'(y)^2} = c.$$

Each of these equations can be easily integrated. Thus

$\arctan f'(x) = cx$ (ignore the additive constant);
$f'(x) = \tan cx;$
$f(x) = -\frac{1}{c} \log |\cos cx|$ (ignore the additive constant).

Likewise,

$$g(y) = \frac{1}{c} \log |\cos cy|.$$

Thus the minimal surface example we obtain has the form

$$z = u(x, y) = \frac{1}{c} \log \left| \frac{\cos cy}{\cos cx} \right|.$$

Example 3 (Catenoid). This is a *surface of revolution* which is also a minimal surface. There are a couple of approaches to this. It is not difficult to assume that $u = u(r)$, where $r = \sqrt{x^2 + y^2}$, derive the corresponding ordinary differential equation from (5′), and then integrate it. The other approach is to regard this as a single-variable problem, where the unknown function $y = u(x) > 0$ is regarded as a curve to be revolved around the x-axis.

The resulting area is then

$$2\pi \int_a^b u\sqrt{1 + u'^2}\,dx.$$

Then the corresponding Euler-Lagrange equation is

$$\frac{\partial}{\partial u}\left(u\sqrt{1+u'^2}\right) - \frac{\partial}{\partial x}\left(\frac{\partial}{\partial u'}(u\sqrt{1+u'^2})\right) = 0.$$

This simplifies to

$$1 + u'^2 - uu'' = 0.$$

This equation is fairly easily integrated, resulting in

$$u(x) = \frac{1}{A}\cosh(Ax+B),$$

where $A > 0$ and B are constants of integration.

B. Geodesics. We just mention this in passing. Given a smooth surface in $\mathbb{R}^3$, there are calculus formulas for the arc length of a curve lying on the surface. The problem of minimizing the arc length of a curve on the given surface which connects two given points is a calculus-of-variations problem. There is a corresponding Euler-Lagrange equation, which is actually a system of two ordinary differentiations, each of second order, in which the unknowns represent the coordinates of a point on the curve.

In the elementary case of the (x,y)-plane itself, a curve given as $x = x(s)$, $y = y(s)$ has length

$$\int_a^b \sqrt{x'^2 + y'^2}\, ds.$$

This does not quite fit the Euler-Lagrange scenario, but the variation idea of Section III leads to an equation

$$\int_a^b \frac{d}{dt}\sqrt{(x'(s)+t\varphi'(s))^2 + (y'(s)+t\psi'(s))^2}\bigg|_{t=0} ds = 0.$$

That is,

$$\int_a^b \frac{x'\varphi' + y'\psi'}{\sqrt{x'^2+y'^2}}\, ds = 0.$$

Here $\varphi(s)$ and $\psi(s)$ are arbitrary, except they are zero at the endpoints. Integrating by parts then gives

$$\int_a^b \left[\left(\frac{-x'}{\sqrt{x'^2+y'^2}}\right)'\varphi + \left(\frac{-y'}{\sqrt{x'^2+y'^2}}\right)'\psi\right] ds = 0.$$

Calculus of Variations

Now we can apply the fundamental lemma to conclude that the Euler-Lagrange equation in this case consists of the *two* equations

$$\left(\frac{x'}{\sqrt{x'^2+y'^2}}\right)' = 0 \quad \text{and} \quad \left(\frac{y'}{\sqrt{x'^2+y'^2}}\right)' = 0.$$

The easiest way to handle these equations is to use the parameter s as the arc-length parameter for the curve, so that

$$x'^2 + y'^2 = 1.$$

Then the Euler-Lagrange equation becomes

$$x'' = 0 \quad \text{and} \quad y'' = 0.$$

That is, x and y are linear functions of s, so that the curve is a straight line.

C. Isoperimetric problem. There is a host of problems of this nature, but we give just one illustration. The example we are going to handle can be stated this way: among all closed curves in $\mathbb{R}^2$ which have a given arc length L, find one which encloses a maximum area A.

This problem has an interesting twist, in that a maximum is sought under a constraining equation. In fact, this is reminiscent of the Lagrange multiplier technique of finite-dimensional calculus. Rather than approach the problem that way, however, we can rig things to fit our pattern. Namely, the quantity $A \div L^2$ is invariant under a change of scale, and therefore that is the function we seek to extremize.

Suppose we parametrize our closed curves in $\mathbb{R}^2$ as $x = x(s)$, $y = y(s)$, $0 \le s \le s_0$. Then from vector calculus we have

$$A = \frac{1}{2}\int_0^{s_0}(xy' - yx')ds \quad \text{(assuming positive orientation)},$$

$$L = \int_0^{s_0}\sqrt{x'^2 + y'^2}ds.$$

Then we want to find the variation in

$$\frac{A}{L^2},$$

using $x(s) + t\varphi(s)$ and $y(s) + t\psi(s)$. We take the derivative of the quotient of the corresponding A and L^2, and then set $t = 0$. We obtain, symbolically,

$$\frac{1}{L^2}\frac{dA}{dt} - \frac{2A}{L^3}\frac{dL}{dt} = 0.$$

Thus, at $t = 0$,

$$\frac{dA}{dt} = \lambda\frac{dL}{dt} \qquad \left(\lambda = \frac{2A}{L}\right).$$

Therefore, just as in the calculations done in section B, we have

$$\frac{1}{2}\int_0^{s_0}(x\psi' + \varphi y' - y\varphi' - \psi x')ds$$
$$= \lambda\int_0^{s_0}\frac{x'\varphi' + y'\psi'}{\sqrt{x'^2 + y'^2}}ds.$$

Integrating by parts produces

$$\frac{1}{2}\int_0^{s_0}(-x'\psi + \varphi y' + y'\varphi - \psi x')ds$$
$$= -\lambda\int_0^{s_0}\left[\left(\frac{x'}{\sqrt{x'^2 + y'^2}}\right)'\varphi + \left(\frac{y'}{\sqrt{x'^2 + y'^2}}\right)'\psi\right]ds.$$

The fundamental lemma now gives

$$-x' = -\lambda\left(\frac{y'}{\sqrt{x'^2 + y'^2}}\right)';$$

$$y' = -\lambda\left(\frac{x'}{\sqrt{x'^2 + y'^2}}\right)'.$$

One integration then gives

$$x - c_1 = \lambda\frac{y'}{\sqrt{x'^2 + y'^2}},$$
$$y - c_2 = -\lambda\frac{x'}{\sqrt{x'^2 + y'^2}}.$$

But then we see that

$$(x - c_1)^2 + (y - c_2)^2 = \lambda^2,$$

so our curve is a circle!

VI. Important disclaimer

We cannot overemphasize that we have not accomplished as much as we would like. In all the above we proceed under the assumption that our problems actually possess solutions. The techniques we have given then enable us to derive significant information about the solutions.

The problem of existence of solutions requires more-or-less sophisticated techniques of geometry and especially analysis. For instance, in our example of geodesics in the plane, it is rather elementary to prove directly that a straight line segment is indeed the unique curve joining two given points. None of what we have discussed is required for such a proof.

To illustrate that some significant analysis may perhaps be required, consider the isoperimetric problem we have just talked about. It would seem that for a general closed curve in the plane, $A \div L^2$ is maximized in the case of a circle, for which $A \div L^2$ equals $\pi r^2 \div (2\pi r)^2 = 1/(4\pi)$. Thus we should imagine that for a general curve in the plane,
$$4\pi A \leq L^2.$$
This is the famous *isoperimetric inequality*. However, you surely realize that we are not even close to proving such an inequality in this talk. The inequality is indeed valid, but an actual proof requires much different kinds of analysis.

In summary, the formal calculus of variations we have talked about leads to very interesting mathematical objects and often paves the way to knowing what to expect to be true.

How Many Equilibria Are There? An Introduction to Morse Theory

Robin Forman

Our goal in this lecture is to investigate a way of counting the equilibria of a dynamical system. Actually, we will not really count the equilibria, but rather we will relate the number of equilibria to the answer of a problem in topology that seems at first glance to have little to do with dynamical systems. The connection between these two subjects was discovered, at least in the form we will present, by Marston Morse, and many of the ideas we will discuss were first introduced in his fundamental works [**Mo1**] and [**Mo2**], but we will also refer to later insights. We warn the reader that this lecture is in the form of an informal discussion. We will make some rather vague comments along the way, with the goal of adding precision as we go along. However, some points will remain imprecise throughout the entire lecture. We hope only to give the reader an appreciation

---

©2004 by the author

for the wonderful subject which now goes under the name of "Morse Theory".

I. Dynamical systems

The first topic on the agenda for today is *dynamical systems*. The sort of dynamical systems we have in mind are those in which a system is acting so as to minimize its energy. Such systems are found in abundance in nature. The equilibrium positions play an important role in our understanding of such a system, because they are the only states of the system that can be observed for more than a fleeting moment. Let us begin our discussion with a simple example.

When I was a student, I would often amuse myself, while listening to a boring lecture, by trying to balance the chair next to mine on its two back legs. (Many friends of mine would balance the chair they were sitting in on the back legs. It is significantly easier to balance a chair on its rear legs if you are sitting in it, because you can shift your own weight to help maintain balance. On the other hand, there is much more at stake.) We all know that there is such an equilibrium position. Now I have a question for you. Why is it that when we enter a room filled with chairs, we never find that some of the chairs are balanced on their back legs. Most of the chairs are resting on all 4 legs, and perhaps we see a few chairs lying on their side, but I have never (yet!) come into a room to find a chair balancing on its back legs.

I'm sure you all know the answer to this mystery. The position of balancing on the two back legs is an *unstable* equilibrium. Even if we did manage to get a chair balancing on its two rear legs, any movement forward or back, no matter how small, would send the chair forward to its standard position, or tipping completely over backwards.

Now let's go further. It seems pretty clear that there is an equilibrium position for the chair in which it is balanced on a single rear leg. Even my daredevil friends never tried that one. Why not? (I am not asking this question in order to encourage you to try!) The answer is that not only is it an unstable equilibrium, it is even more unstable than the equilibrium position of balancing on two legs.

How Many Equilibria?

What does it mean for one unstable equilibrium to be more unstable than another? How can we quantify levels of instability? Before reading further the reader should think about this point on his or her own.

Okay, perhaps that's long enough. I will now describe one way of measuring the instability of an equilibrium point. If we are balancing a chair on its two rear legs, there is only one component of motion, the forward-back component, that we need to worry about. If the chair is balancing on its two rear legs, unless there is a major disruption, it is unlikely to tip over on its side. We say that the forward-back component is an *unstable component* of direction. On the other hand, if we are balancing a chair on a single rear leg, we have to worry about two components of motion, the forward-back component, and the left-right component, since we can easily tip over in any direction. (Fortunately, with the combined help of gravity and the floor, we do not have to worry too much about the up-down component.) In this case there are two unstable components of direction.

With all this in mind, let us define the *index* of an equilibrium position to be the number of unstable directional components. For example, the index of any stable equilibrium is 0. As we said earlier, our goal is to investigate the number of equilibria of a dynamical system. One of Morse's great insights is that the problem becomes easier if we keep track of the index of each of the equilibria.

Before moving on, let us spend a bit more time exploring the concept of the index of an equilibrium point. It is important that we come to grips with this idea, because it is central to the subject. Suppose there is a ball rolling around on the one-dimensional landscape shown in Figure 1. In this case, the ball is acting so as to minimize its height (or equivalently, its potential energy due to gravity). The equilibria are those points such that if we placed the ball there and let go (so that the ball starts with zero velocity) it will stay there. Another way to say this is that they are precisely the points such that the tangent line to the landscape is horizontal. (Throughout this discussion we will assume that the height function is differentiable, so that we may speak about tangent lines. Soon we will also assume that the second derivative of the height function exists.) There are three such points, which we have labeled A, B and C. The point A is a stable equilibrium, since if the ball begins at a point near A, it will roll towards A. Therefore, point A has index 0. The point B is clearly unstable, since if the ball begins near the point B, it will roll away from B. This equilibrium has index 1, since it is unstable in the left-right direction, and there are no other directions available to the ball. How about the point C? It is stable to the left (since if the ball starts slightly to the left of the point C it will roll towards C), but unstable to the right (since if a ball is placed slightly to the right of C it will roll away from C). What is the index of C? Should we say it is stable or unstable in the left-right component of the direction? There does not seem to be any correct answer to this question, so we will simply say that the index of the point C is undefined.

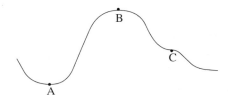

Figure 1. A one-dimensional landscape with three equilibria

I'm sure that this last paragraph reminded you a bit of some discussions you had in your first calculus course. The point A is a local minimum of the energy (= height) function, the point B is a local maximum, and C is an inflection point. In fact, this entire discussion is probably best carried out in the language of calculus. The equilibria are those points where the tangent line to the graph of the height function is horizontal. Those are precisely the critical points of the height function. Now suppose that p is a critical point of the energy function E, so that $E'(p) = 0$. If, in addition, $E''(p) > 0$, then we know that p is a local minimum of E, so p is a stable equilibrium, and hence has index 0. If $E''(p) < 0$, then we know that p is a local maximum of E, so p is an unstable equilibrium of index 1. If $E''(p) = 0$, then the second derivative test does not tell us the index of p, or even if the index is well-defined. If $E''(p) = 0$ we say that p is

How Many Equilibria?

a *degenerate critical point of E*. Conversely, if $E''(p) \neq 0$ we say that p is a *nondegenerate critical point of E*. If all of the critical points of E are nondegenerate, so that, in particular, all of the critical points have a well-defined index that can be determined from the second derivative test, then we say that the energy function E is a *Morse function*.

In the example of a ball moving along the landscape, the set of possible positions of the ball can be identified with the points on the x-axis (i.e., if you tell me the x-coordinate of the ball, I know immediately where it is). The set of possible positions of a dynamical system is called the *configuration space* of the system. In the previous example, the configuration space can be identified with a one-dimensional line. We will now consider dynamical systems with more interesting configuration spaces.

Example 1: Suppose we have a pendulum on a rigid rod. The configuration space for this system, that is, the set of possible positions of the pendulum, can be identified with a circle (see Figure 2). The equilibria are the points labeled A and B. The point A is a stable equilibrium and has index 0, and the point B is an unstable equilibrium and has index 1. We will record this information as follows:

Configuration space = circle

Number of equilibria of index $0 = 1$

Number of equilibria of index $1 = 1$

Number of equilibria of index $\geq 2 = 0$.

Our discussion can also be carried out in higher dimensions. Suppose we have a differentiable function $E : R^2 \to R$. We can graph this function in R^3 and think of a ball rolling around on this landscape. The equilibria are the points where the tangent plane is horizontal, and these are precisely the critical points of the function E. In our multi-variable calculus course, we are taught the second derivative test for such functions. If p is a critical point of a function E, we

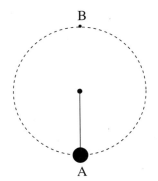

Figure 2. A pendulum on a rigid rod

consider the Hessian of E at p. This is the 2×2 matrix of second derivatives

$$\begin{pmatrix} \frac{\partial^2 E(p)}{\partial x^2} & \frac{\partial^2 E(p)}{\partial x \partial y} \\ \frac{\partial^2 E(p)}{\partial x \partial y} & \frac{\partial^2 E(p)}{\partial y^2} \end{pmatrix}.$$

If both eigenvalues of this matrix are positive, for example if $E(x,y) = x^2 + y^2$, and p is the origin, then p is a local minimum of the energy function (Figure 3(i)). This implies that p is a stable equilibrium and hence has index 0. If the Hessian has two negative eigenvalues, for example if $E(x,y) = -x^2 - y^2$, and again p is the origin, then p is a local maximum of the energy function (Figure 3(ii)). This implies that at p both the x and the y components of direction are unstable components, so p is an equilibrium of index 2. If the Hessian has one positive eigenvalue and one negative eigenvalue, for example if $E(x,y) = -x^2 + y^2$, and again p is the origin, then p is a saddle point (Figure 3(iii)). In this case, the x component of direction (indicated by the dotted curve through p in Figure 3(iii)) is an unstable component, since if we place a ball at p and then move it slightly in the x direction and let go, it will roll away from p. On the other hand, p is stable in the y direction (the solid curve through the point p in Figure 3(iii)), since if we place a ball at p and then move it slightly in the y direction and let go, it will roll towards p. Therefore, there is precisely

How Many Equilibria?

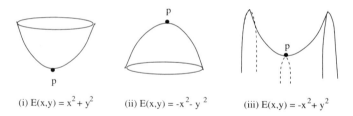

(i) $E(x,y) = x^2 + y^2$ (ii) $E(x,y) = -x^2 - y^2$ (iii) $E(x,y) = -x^2 + y^2$

Figure 3. Examples of two-dimensional equilibria

one unstable component of direction at p, so p is an equilibrium of index 1.

If the Hessian has a zero eigenvalue, then the second derivative test does not tell us the index of p, or even whether p has a well-defined index. In this case, just as in the one-dimensional setting, we say that p is a degenerate critical point.

This discussion can be carried out in any dimension. Suppose that the configuration space is m-dimensional, and that near a critical point p the energy function E has the form

(1) $\quad E(x_1, x_2, \ldots, x_m) = -x_1^2 - x_2^2 - x_3^2 - \cdots - x_i^2 + x_{i+1}^2 + \cdots + x_m^2,$

so that there are exactly i unstable components of direction (i.e., directions in which the energy is decreasing). Then the critical point p has index i. Just as in the case of 2 dimensions, we can use the second derivative test to detect the index of a critical point.

The second derivative test: Suppose that p is a critical point of an energy function E, and the Hessian of E at p has no zero eigenvalues. Then the index of E at p is precisely the number of negative eigenvalues of the Hessian.

I have not labeled this statement a theorem because we do not yet have a precise definition for the index. In fact, this statement is often used to define the index of an equilibrium point. An alternate point of view is that one can define a critical point p to have index i if near p the energy function E looks like the function shown in (1). If one takes this as the definition for the index, then one has to prove the above statement by showing that if the Hessian of E at p has i negative eigenvalues (and no zero eigenvalues), then E looks like the function (1). Here, we have been using a rather vague phrase "looks like" in reference to functions. Unfortunately, I do not think that it would be worth the effort to make this phrase precise. For a precise statement of what we mean for a function to look like another in this context, and a proof of the second derivative test stated above, see "The Morse Lemma" (the name given to this second derivative test), Lemma 2.2 in [**Mi1**].

Now let us apply these ideas to some 2-dimensional examples.

Example 2: Suppose we start with a round metal globe. The mathematical name for this shape is a *sphere*, or a *2-sphere* if we wish to emphasize that it is 2-dimensional, and is often denoted by the symbol S^2. Now suppose we place a marble on its surface, and that the marble is magnetized so that it stays on the globe as it rolls around. The equilibria are those points with the property that if the marble is placed there with zero velocity, then the marble will not move. These are the points where the tangent plane is horizontal. There are 2 equilibrium points on the sphere (Figure 4), which we have labeled A and B, and they clearly have indices 0 and 2, respectively.

Configuration space = sphere

Number of equilibria of index $0 = 1$

Number of equilibria of index $1 = 0$

Number of equilibria of index $2 = 1$

Number of equilibria of index $\geq 3 = 0$.

Example 3: This time we start with a metal inner tube. The mathematical name for this shape is a *torus*. Now suppose we sit the metal torus on one end (see Figure 5) and place a magnetized marble on its surface. Here there are 4 equilibrium points, which we have labeled A, B, C and D. The point A is a stable equilibrium, and has index 0.

Figure 4. Equilibrium points for a magnetized marble rolling around on a metal globe

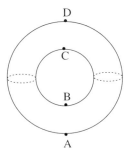

Figure 5. Equilibrium points for a magnetized marble rolling around on a metal torus

The point D is a local maximum of the energy function and has index 2. The points B and C are a bit trickier. In fact, if one considers a small piece of the torus near the point B, it looks very much like the saddle point we drew in Figure 3(iii). The point B is stable in the left-right component of direction, and unstable in the forward-back component. Therefore B has index 1. The same holds true for the point C except that C is unstable in the left-right component of direction, and stable in the forward-back direction. Therefore, the point C also has index 1.

Configuration space = torus

Number of equilibria of index 0 = 1

Number of equilibria of index 1 = 2

Number of equilibria of index 2 = 1

Number of equilibria of index $\geq 3 = 0$.

II. Topology

We will now briefly leave the subject of dynamical systems and begin a discussion of some topics in topology. You probably already know something about topology. It is sometimes called "rubber geometry" because we will say that two shapes are topologically the same if one can be made into the other by stretching, pulling and twisting, and other similar operations. No cutting or pasting is allowed. If two shapes are the same in this way, we will say they are *topologically equivalent*, or (using the fancy word) *homeomorphic* (homeo = same, morph = shape). The sort of question we will be asking is typical in all branches of science (and, in fact, in many other areas of human endeavor). We will begin by describing a collection of simple objects, the building blocks of the theory. The main problem will then be to investigate how more complicated objects can be built from these building blocks.

The building blocks in the theory we wish to discuss are called *cells*, and there is one cell for each dimension.

The 0-dimensional cells are points (also called vertices).

The 1-dimensional cells are open intervals.

The 2-dimensional cells are discs (without their boundary). Since we are working in the world of topology, we can stretch these discs, and

How Many Equilibria? 23

we can draw them as squares, or triangles, if we like. They are all 2-dimensional cells.

The 3-dimensional cells are the inside of a cube, or the inside of a sphere, or the inside of a pyramid, or ...

We illustrate these cells in Figure 6. The dotted lines in this figure are meant to indicate that the cells do not contain the points on their boundary.

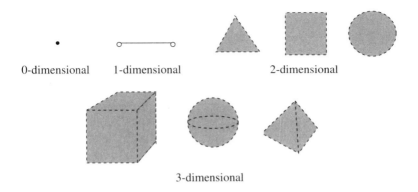

Figure 6. Topological cells

We can continue, and define cells of every dimension, but we will stop here. It is useful to introduce a bit of math shorthand. Instead of writing out "0-dimensional cell" each time, it is traditional to refer simply to a "0-cell". Similarly, we will sometimes refer to a 1-cell, or 2-cell.

Now let us return to the basic question. If we are given a shape, how can we build it out of these building blocks?

Example 1: Let's try this in the case of the circle (see Figure 7). It is pretty easy to see that if I remove a single point (i.e., a 0-dimensional cell) from the circle, then what remains is (topologically) an open interval (i.e., a 1-dimensional cell). Therefore, the circle can be built from one 0-dimensional cell and one 1-dimensional cell. Let us summarize this as follows.

Topological space = circle

Number of cells of dimension 0 = 1

Number of cells of dimension 1 = 1

Number of cells of dimension ≥ 2 = 0.

Figure 7. A decomposition of a circle into cells

Example 2: For our next example, we consider the 2-dimensional sphere (see Figure 8). If we remove a single point from the sphere, what remains can be stretched out flat to a round disc (i.e., a 2-dimensional cell). Therefore, the 2-sphere can be built from one 0-dimensional cell and one 2-dimensional cell.

Topological space = sphere

Number of cells of dimension 0 = 1

Number of cells of dimension 1 = 0

Number of cells of dimension 2 = 1

Number of cells of dimension ≥ 3 = 0.

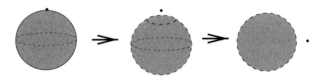

Figure 8. A decomposition of a 2-sphere into cells

How Many Equilibria?

Example 3: This time we start with a torus. We can build a torus as follows (see Figure 9). If we take a circle out of the torus, what remains is a cylinder which does not contain its boundary circles. We can build this circle from one 0-cell and one 1-cell. We can build the cylinder from one 1-cell and one 2-cell. Putting these constructions together, we have just shown how to build the torus from one 0-cell, two 1-cells, and one 2-cell.

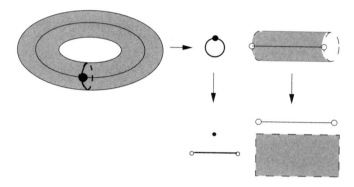

Figure 9. Decomposing the torus into cells

Topological space = torus

Number of cells of dimension 0 = 1

Number of cells of dimension 1 = 2

Number of cells of dimension 2 = 1

Number of cells of dimension $\geq 3 = 0$.

III. Morse Theory

By now I hope the reader will have noticed the remarkable similarity between the examples in Section I and the examples in Section II. We first observe that the shapes considered in Section II are precisely the configuration spaces of the dynamical systems considered in Section I. Moreover, the number of cells required to build the configuration space seems to be the same as the number of equilibria of the dynamical system, with the dimension of the cell corresponding to the index of the equilibrium point. The main theorem of this lecture is that this is a general phenomenon. We are going to state the theorem using the precise mathematical terminology, all of which will be explained in the remainder of this section. What follows is the main theorem of Morse Theory.

Theorem: Let M be a closed, compact, smooth submanifold of Euclidean space (of any dimension). Let $E : M \to R$ be a smooth, real-valued function on M. Suppose that every critical point of E is nondegenerate. Then M can be built from a finite collection of cells, with exactly one cell of dimension i for each critical point of index i.

Some of the words in this theorem need to be explained. I will not give precise definitions, but only a rather vague description of what the hypotheses are requiring. First note that the theorem refers to *Euclidean space (of any dimension)*. All of the examples we have considered take place in R^2 or R^3, but the phrase in the theorem means that everything can be placed in R^k for any k. A subset of a Euclidean space R^k is a *smooth submanifold* if it has the property that for each point p in the subset, the set of points in the subset which are near p looks just like the set of all points near the origin in some Euclidean space. In Figure 10 we show three subsets of R^2 which are *not* smooth submanifolds because in each case the point labeled A does not satisfy this condition. Note that in Figure 10(iii), near the point A the subset looks like the set of points near the origin in R^1 in a topological sense, since in topology we can always straighten out corners, but we actually need the space to look like Euclidean space in a stronger differentiable sense so that we can make sense of the idea of taking derivatives of functions on our space. In fact, we will need somewhat more because we will need to take the second derivative of our functions on M.

We must now explain the meaning of the word *compact*. A subset of Euclidean space is compact if and only if it satisfies two other

How Many Equilibria? 27

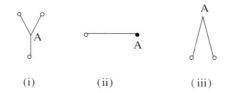

(i)   (ii)   (iii)

Figure 10. Subsets of R^2 which are not smooth submanifolds

properties. Namely, it must be *closed* and *bounded*. A subset M of R^k is closed if it has the following property. Suppose there is a point q in R^k such that we can get as close to q as we want while staying in M. Then q is required to be in M. For example, suppose we let M be the subset of R^1 consisting of all of R^1 except the number 0. Then M is not closed because you can get as close to 0 as you like while staying in M, but the point 0 itself is not in M. The same problem occurs if M is any open interval, for example, if $M = (0,1)$. Then one can get as close as we like to the numbers 0 and 1 while staying in M, but the numbers 0 and 1 are not themselves in M. The condition that M is bounded means simply that it is possible to surround M by a (possibly very large) ball in R^k, i.e., M does not go off to infinity in any directions. For example, the x-axis sitting in R^2 is closed but not compact.

Note that the three examples we examined in Sections 1 and 2 satisfied the hypotheses. It is a very good exercise for the readers to think about what can go wrong with the theorem if M is not required to be compact.

I will leave the reader with one last note concerning the statement of the theorem. We required that M be a subset of some Euclidean space. In fact, that is irrelevant for the theorem. We do need M to be a space on which it makes sense to take derivatives of functions. Such spaces are called *smooth manifolds*. I stated the theorem in this manner only because abstract manifolds, that is, those which exist on their own, without reference to a surrounding Euclidean space, are somewhat harder to describe and to imagine.

IV. What the main theorem does not say

There is occasionally some confusion as to what the main theorem actually says. It does not say, for example, that all energy functions have the same number of critical points, or that the cell complex that you get in this manner is necessarily the "best possible" (i.e., has the fewest number of cells). Consider the example shown in Figure 11(i). In this case, our space is a topological circle. We see that there are four critical points, labeled A, B, C and D, with A and C having index 0, and B and D having index 1. Therefore, the main theorem implies that the circle can be built from two 0-cells and two 1-cells. This is shown in Figure 11(ii). The reader should compare this example with Example 1 of Sections I and II.

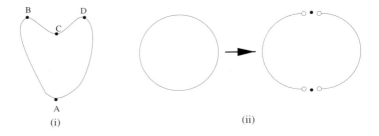

Figure 11. A configuration space which is a topological circle

V. The idea of the proof

The proof we are going to describe is due to Smale [**Sm1**]. I apologize in advance that some details will be imprecise, and others will be precise but will have unexplained terms. Still I hope that the reader will walk away with some understanding of what is going on.

Suppose p is a critical point of index i. Draw a small piece of the unstable directions near p. After adding the point p to this set, it will be a small i-cell. This is true because a critical point has index i precisely when there is an i-dimensional family of unstable directions. See Figure 12(i) where we have done this for each of the critical points of the function on the torus considered in Example 3 of Section I.

How Many Equilibria?

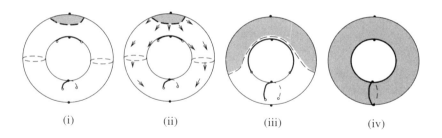

(i) (ii) (iii) (iv)

Figure 12. Growing a cell decomposition on the torus

The next step is to draw a little arrow at every point on the manifold, indicating the direction in which the energy is decreasing the fastest (Figure 12(ii)). You may know that the *gradient* of the energy function points in the direction in which the energy is increasing the fastest, so the arrow we are drawing points precisely in the opposite direction of the gradient. Now think of each point on the manifold flowing in the direction of the arrows we drew. If you are not sure what this means, think of putting a drop of molasses at a point on the manifold and watching it slowly make its way down. The drop will come very close to flowing in the direction of the arrows we have drawn.

Now watch what is happening to each of the cells we drew. As they flow they grow bigger and bigger, but they remain cells of the appropriate dimension (Figure 12(iii)). The key point is that as we let more and more time go by, these cells fill up more and more of the manifold. If we let the amount of time go to infinity (!), the cells fill up the entire manifold (Figure 12(iv)), and this shows that the manifold can be decomposed into the desired number of cells.

How do we know that the cells will fill up the entire manifold? Let q be a point in M. We need to see that one of the cells, which we are watching grow larger and larger, will eventually contain q. Which cell will it be? There is a fun way to tell. Let a drop of molasses begin at the point q, and this time let it flow **up** the manifold (or equivalently, turn the manifold upside down and let the drop flow down). If we let time approach infinity, the drop will get closer and closer to a critical point (here we are definitely using the compactness of M). The critical point the drop approaches is precisely the critical point whose cell will eventually contain q.

This point of view enables us to give a concise description of the resulting decomposition of M into cells. For each critical point p, let $U(p)$ denote the set of all points in M which, when flowing **up** the manifold, flow towards p. We will include the point p in $U(p)$ even though it doesn't flow at all. The set $U(p)$ is called the *unstable cell* associated to p (or sometimes the *descending cell* associated to p). The crucial observations are:

1) $U(p)$ is a cell, and its dimension is the index of p.

2) If p and q are distinct critical points, then $U(p)$ and $U(q)$ are disjoint.

3) M is equal to the union of all the $U(p)$, where the union is taken over all critical points p.

VI. What does it really mean to build a shape from cells?

So far we have been a bit cavalier with our language when it comes to "building shapes out of cells". In fact, the sort of decomposition into cells that is provided by Morse Theory has some special properties, which we will now discuss.

Suppose we already have a shape, and we wish to add a 1-cell to it. We know a 1-cell is an open interval, so that is what we are adding to our space. However, I will now add an important restriction as to how that 1-cell can be added. From now on, we are not permitted to do this in any way we please. We must fill in the 2 endpoints of the interval with points that are already in our space. In other words, we begin with a closed interval, and glue the 2 endpoints to points in our space, so that the only points we are adding to the space are those in the open interval. It is important that we glue both points to our original space. For example, in Figure 13 we illustrate the case in which we begin with a 0-cell. In Figure 13(i) we show how to add a 1-cell to the 0-cell. The drawing in Figure 13(ii) is not permitted.

How Many Equilibria?

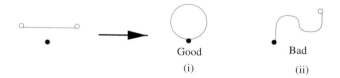

Figure 13. Adding a 1-cell to a point

Adding a 2-dimensional cell is defined similarly. Instead of starting with a 2-dimensional disc without its boundary, we think instead of the 2-dimensional disc with its boundary. The boundary is just a topological circle. To add a 2-cell to a space, we must glue every point on that circle to a point in our space. Moreover, we must do this in a continuous way.

Adding an i-cell, for $i > 2$ is defined similarly. A *cell complex* is a space that can be constructed by starting with a single point and adding one cell at a time in this manner.

The decomposition of M into cells which is provided by the Morse Theorem has the important property of giving M the structure of a cell complex.

Note that to give a space the structure of a cell complex, one must start with a single point (which we count as a 0-cell) and then order the remaining cells, i.e., declare which one to add first, which one to add next, etc. How does this arise for M? Simply order the critical points of the energy function according to the value at the energy function at that point (ties can be broken arbitrarily). The first critical point in line will be the minimizer of the energy. This is always a stable critical point, and hence has index 0. This gives us our 0-cell to start with. We then add the cells one at a time, according to the value of the energy function at the critical point. In Figure 14 we illustrate this process for the example of the torus with the height function shown in Figure 4. In this case we add the cells corresponding to the critical points A, B, C, D in that order.

We can now state a more precise version of the main theorem of Morse Theory.

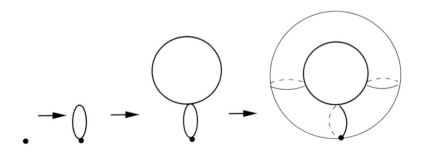

Figure 14. The torus as a cell complex

Theorem: Let M be a smooth compact submanifold of Euclidean space (of any dimension). Let $E : M \to R$ be a smooth, real-valued function on M. Suppose that every critical point of E is nondegenerate. Then M is homeomorphic (i.e., topologically equivalent) to a cell complex which has exactly one cell of dimension i for each critical point of index i.

There is one last subtlety to discuss. There is not complete uniformity in the literature as to the definition of a cell complex. As it is often defined, a cell complex is required to satisfy an additional condition. We required that when we add a cell to a space, the entire boundary of the cell must be glued to the space. There is often an additional requirement. Recall that the space we are adding the cell to is itself a union of cells. The additional requirement is that the boundary of an i-cell can only be glued to cells of dimension less than i.

Take a look at the construction of the torus as a cell complex shown in Figure 14. This is not a cell-complex in this new, more restrictive, sense, because the boundary of the second 1-cell we add is glued to the first 1-cell, and this is not permitted. However, if we just tilt the torus slightly before beginning the growing process shown in Figure 12, then the second 1-cell would miss the first 1-cell entirely, and the result would be a cell-complex in our new sense. This always works. Given a Morse function E, if E does not give rise to a good cell-complex, then a generic small perturbation of E will

How Many Equilibria?

do the trick. (The word "generic" means that those perturbations which will not work form a very small set.) Smale was the first to investigate this issue, and energy functions which give rise to good cell decompositions are now called Morse-Smale functions. (In fact, somewhat more is required of a function before it is called Morse-Smale. Not only must the boundary of an i-cell only meet cells of lower dimension, but they must meet in a nice way. I will not say any more about this. See [**Sm1**] for details.)

VII. What now?

If all there were to Morse Theory is the main theorem we have discussed, then I would still think it was a beautiful subject, but I would probably not be making such a fuss about it. In fact, Morse's work led to a veritable revolution in the study of the topology of smooth manifolds. There is simply no time to give an overview of the hundreds of applications of Morse Theory which have appeared in the literature. There are many examples of results that have been proved rather easily using Morse Theory and yet are quite difficult to prove by other means. The book [**Mi1**] describes some of these applications. Here, I will simply briefly describe two very striking (and historically significant) applications. I know that my brief remarks will be insufficient to give the reader a true understanding of these great works. I do hope that perhaps the reader will be sufficiently intrigued to study these topics further.

We recall that the main theorem of Morse Theory relates the number of equilibrium points of a dynamical system to the number of cells required to build the configuration space. One of the wonderful aspects of this theorem is that it has been very powerfully applied in each direction. That is, sometimes one wishes to understand the number of equilibria of a dynamical system, so one studies the topology of the configuration space and then applies Morse Theory. On the other hand, sometimes one wants to understand the topology of a space. One method is to put an energy function on the space. Information about the critical points of this function can then be translated into information about the original space.

One of the first major applications of Morse Theory was Morse's investigation of the number of geodesics between two points in a manifold (see page 248 in [**Mo2**] and Part III of [**Mi1**]). Very roughly speaking, suppose we stretch a rubber string between two points in a manifold and glue the endpoints of the string to those two points. If, when we let go of the rubber string it stays exactly as it is (we are assuming that it is restricted to stay on the manifold), then the path followed by the rubber string is a geodesic (see, for example, Part II of [**Mi1**] for a precise definition). The rubber string moves so as to minimize its energy, so this set-up is ripe for an application of Morse Theory. Namely, the configuration space of our dynamical system is the space of all possible ways for a rubber string to go from one point to another, i.e., the set of all paths from one point to another. The geodesics are precisely the equilibria of this dynamical system. Morse studied the number of geodesics by investigating the topology of the configuration space and then applying his theory. For example, he was able to deduce that there are always infinitely many geodesics between any two points in a smooth compact manifold. (We have something more to say about this example a bit later.)

In Morse's work on geodesics, information about the topology of the configuration space was used to deduce information about the number of equilibria of the system. We now describe an example in which the flow of information goes the other way. In [**Sm2**] Smale used Morse Theory to prove the higher-dimensional Poincaré conjecture. The Poincaré conjecture states that any manifold which "looks like" a sphere, in some weak topological sense, is, in fact, topologically equivalent to a sphere (unfortunately, there is no time to explain the conjecture in any more detail than that). Smale's method was to put an energy function on the manifold and to then study the equilibria of the resulting dynamical system. He proved that if a manifold (of dimension ≥ 5) satisfies the hypothesis, then there is an energy function which has only two critical points, namely the maximum and the minimum. It then follows from a theorem of Reeb's that the manifold is a sphere (see Theorem 4.1 of [**Mi1**]). While Smale's proof used Morse Theory quite extensively, he also used techniques from an approach to topology known as "handlebody theory". In [**Mi2**]

Milnor presents a proof which is entirely in the language of Morse Theory.

A great place to read about Morse Theory, as well as some of the earlier, exciting, applications, is Milnor's wonderful book [**Mi1**]. The reader should be warned, however, that this book, like Morse's early writings, does not take the same point of view we have chosen, so some parts of the discussion may not look very familiar. Our discussion is more in line with the philosophy of [**Sm1**]. In addition to Morse, Milnor and Smale, Raoul Bott is one of the great practitioners of Morse Theory, and the reader should certainly take a look at his beautiful survey article [**Bo**] for a look at some of the recent developments in the theory.

Morse Theory has been extended and generalized far beyond what we presented in this lecture. Again, we will have to be content with just two examples. Some problems concerning the topology of spaces which are not smooth manifolds have been successfully solved by developing versions of Morse Theory that can be applied to more general spaces (see, for example, [**GM**] and [**Fo**]). Perhaps the most exciting development has been the application of Morse Theory to infinite-dimensional manifolds (!). It is interesting to note that one of Morse's first applications of his theory, which we described above, was to the study of an infinite-dimensional manifold, the space of all paths connecting two points in a (finite-dimensional) manifold. However, he studied this space by considering finite-dimensional approximations to the space, and applying the finite-dimensional theory. Now we have the know-how to study the infinite-dimensional space of paths directly (see, for example, [**Pa**], [**Sm3**]). More recently, Floer's application of ideas from Morse Theory to some infinite-dimensional manifolds ([**Fl**]) has resulted in some extremely important advances in mathematical physics and topology.

I must warn you that this lecture is not sufficient preparation for reading most of the papers referenced in the previous paragraphs. However, I wanted to mention them so that you could see that Morse Theory has been a growing and vibrant subject ever since its introduction almost 75 years ago. I think that it is one of the most beautiful mathematical developments of the last century. The study of Morse

Theory has given me a lot of joy, and I am happy to have had this occasion to share the subject with you.

References

[Bo] R. Bott, *Morse Theory Indomitable*, Publ. Math. I.H.E.S. **68** (1988), 99–117.

[Fl] A. Floer, *An instanton-invariant for 3-manifolds*, Comm. Math. Phys. **118** (1988), 215–240.

[Fo] R. Forman, *Morse Theory for Cell Complexes*, Adv. in Math. **134** (1998), 90–145.

[GM] M. Goresky and R. MacPherson, *Stratified Morse Theory*, in Singularities, Part I (Arcata, CA, 1981), Proc. Sympos. Pure Math., 40, Amer. Math. Soc., R.I., (1983), 517-533.

[Mi1] J. Milnor, *Morse Theory*, Annals of Mathematics Studies No. 51, Princeton University Press, 1962.

[Mi2] _____, *Lectures on the h-Cobordism Theorem*, Princeton Mathematical Notes, Princeton University Press, 1965.

[Mo1] M. Morse, *Relations Between the Critical Points of a Real Function of n Independent Variables*, Trans. Amer. Math. Soc. **27** (1925), 345–396.

[Mo2] M. Morse, *The Calculus of Variations in the Large*, American Mathematical Society Colloquium Publications, vol. 18, American Mathematical Society, Providence, R.I., (1934).

[Pa] R. Palais, *Morse Theory on Hilbert Manifolds*, Topology **2** (1963), 299-340.

[Sm1] S. Smale, *On Gradient Dynamical Systems*, Annals of Math. **74** (1961), 199–206.

[Sm2] _____, *The Generalized Poincaré Conjecture in Dimensions Greater than Four*, Annals of Math. **74** (1961), 391–406.

[Sm3] _____, *Morse Theory and a Non-linear Generalization of the Dirichlet Problem*, Annals of Math. (2) **80** (1964), 382–396.

Aye, There's the Rub. An Inquiry into Why a Plucked String Comes to Rest

Steven J. Cox

1. Introduction

Newton's first law implies that a plucked string will remain in motion unless impeded by some additional force. Experience teaches that the energy in the average pluck of a guitar string is dissipated within 10 to 20 seconds. Where does this energy go? Does the string rub against itself and/or against its environment? I wish to identify the nature of this dissipation. I take the broad view – to identify is to construct a mathematical model that is physically plausible and yields a reasonable fit to experimental data.

Our first task will be to gather experimental data. In §2 we measure the time and frequency response of a plucked string. We build a preliminary model in §3 via the Principle of Least Action. We augment this in §4 arriving at the *damped wave equation*. We

©2004 by the author

determine, numerically, that the dissipative force that best matches experimental data is concentrated at the string's ends, where it rubs against its supports. As further demonstration, we place a magnetic damper at the string's midpoint and show that our technique detects both its strength and position. Along the way we shall invoke the Discrete Fourier Transform, the method of Least Squares, the Calculus of Variations, and the solution of partial differential equations via eigenfunction expansions.

2. Acquiring the data

Regarding equipment, I have followed, to a large extent, the advice of Lord Rayleigh [R, Vol.1, §125]:

> "For quantitative investigations into the laws of strings, the sonometer is employed. By means of a weight hanging over a pulley, a catgut, or a metallic wire, is stretched across two bridges mounted on a resonance case. A movable bridge, whose position is estimated by a scale running parallel to the wire, gives the means of shortening the efficient portion of the wire to any desired extent. The vibrations may be excited by plucking, as in the harp, or with a bow (well supplied with rosin), as in the violin."

Rayleigh proceeded to estimate the string's natural frequencies by aural comparison with struck tuning forks of known natural frequency. I adopt the sonometer but substitute for ear and fork an electromagnetic pick-up, analog-to-digital converter, and the Discrete Fourier Transform. For hardware, with regard to the photograph in Figure 1, I have used the WA–9611 Sonometer (long object in foreground) and the WA–9613 Detector Coil (small black box positioned under the midpoint of the string) of PASCO Scientific and an AT–MIO–16E Data Acquisition Card from National Instruments. The detector coil returns a voltage that is proportional to the time rate of change of the coil's magnetic flux, which in turn is proportional to the velocity of the string in a neighborhood of the detector.

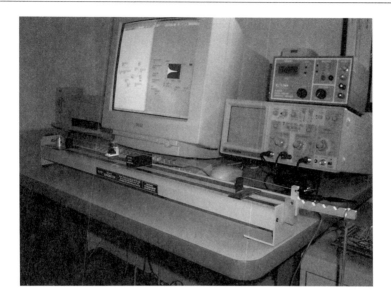

Figure 1. The experimental apparatus

The chosen string possessed a uniform linear density of

(2.1) $$\rho = 0.0015 \, \text{kg/m}.$$

The distance between the two black posts provided an effective length of

(2.2) $$\ell = 0.6 \, \text{m}.$$

Finally, suspending 0.55 kg from the lever in the right corner produced a tension of

(2.3) $$\tau = 26.95 \, \text{m kg/s}^2.$$

The small Lego object to the left of the detector is the magnetic damper that I used in the experiment of §6. In a typical experiment I would pluck the string, in the absence of the magnetic damper, and sample the voltage, 10000 times per second, for about 15 seconds. One would need a page at least 18 inches wide in order to produce a meaningful plot of such a time series. We content ourselves therefore with a plot of every tenth sample.

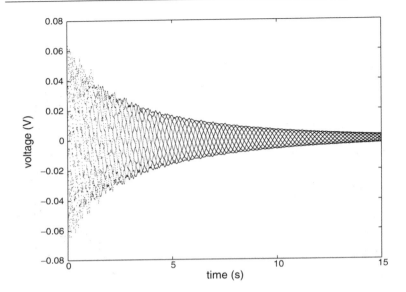

Figure 2. Response of a plucked string

This signal appears, at least eventually, to be composed of a sum of decaying sinusoids. The Discrete Fourier Transform is an excellent tool for extracting their respective frequencies. If v denotes the vector of $N = 150000$ voltages depicted in Figure 2, then the Discrete Fourier Transform of v is

$$v_k^\# = \sum_{n=1}^{N} v_n \exp(-2\pi i (k-1)(n-1)/N), \quad 1 \le k \le N.$$

The relationship between $v^\#$ and the Fourier coefficients a and b in

$$(2.4) \quad v(n) = a_0 + \sum_{k=1}^{N/2} a_k \cos\left(2\pi \frac{k}{Nh} t(n)\right) + b_k \sin\left(2\pi \frac{k}{Nh} t(n)\right)$$

is

$$a_0 = \frac{1}{N} v_1^\#, \quad a_k = \frac{2}{N} \Re(v_{k+1}^\#), \quad \text{and} \quad b_k = -\frac{2}{N} \Im(v_{k+1}^\#),$$

where $t(n)$ is the time at the nth sample and $h = 0.0001$ is the time (in seconds) between samples. We have used $\Re$ and $\Im$ to denote the real and imaginary parts respectively. In light of (2.4) we plot the natural

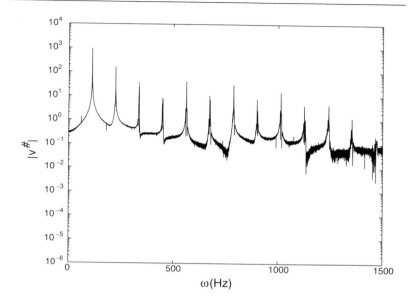

Figure 3. Magnitude of the Discrete Fourier Transform of the signal in Figure 2

logarithm of the magnitude of $v_k^{\#}$ (computed via the `fft` command in Matlab) versus the frequency

$$\omega_k = \frac{k}{Nh}.$$

(See Figure 3.)

One sees immediately that the energy in v is concentrated at integer multiples of about 111 Hz. We speak of these as the resonant (or natural) frequencies of the string. In order to discern at what rate(s) the associated resonant modes decay we attempt to fit our voltage readings to a function of the form

(2.5) $$\phi(t;p) = \sum_{j=1}^{m} p_{j,1} \exp(p_{j,2}t) \cos(2\pi p_{j,3}t + p_{j,4})$$

parametrized by the m-by-4 matrix p. By 'fit' we mean to choose p in order that the sum, over all samples, of the squares of the differences

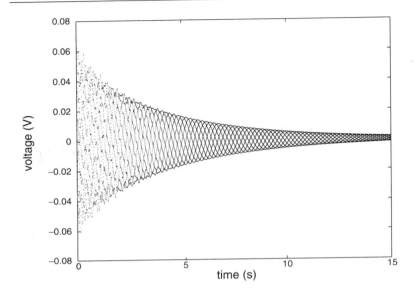

Figure 4. Best fit, in the sense of (2.1), to Figure 2

is minimized. That is, we choose the p that solves

$$(2.6) \qquad \min_{p \in \mathbf{R}^{4m}} \sum_{n=1}^{N} |v_n - \phi(nh; p)|^2.$$

This problem is readily solved in Matlab with a call to `leastsq`. In particular, asking for $m = 8$ 4–tuples, Matlab returned

$$(2.7) \qquad p = \begin{pmatrix} -0.0485 & -0.2181 & 111.6934 & -3.5923 \\ -0.0149 & -0.4956 & 223.3987 & -2.5630 \\ -0.0038 & -0.7865 & 335.1198 & -1.5076 \\ -0.0010 & -1.0670 & 446.8129 & -0.2750 \\ 0.0050 & -0.5977 & 562.1369 & -1.8865 \\ 0.0019 & -0.9026 & 673.8576 & -0.8854 \\ 0.0050 & -0.8177 & 788.0784 & -1.1569 \\ 0.0017 & -1.0837 & 899.7653 & 0.0524 \end{pmatrix}$$

and a value of 0.333 for (2.6). We recognize in the third column of p the aforementioned resonant frequencies. The second column lists their respective decay rates. We wish to point out that the higher frequencies decay 4 to 5 times faster than the lowest. Although p has indeed captured the correct resonant frequencies, you may wonder what the associated ϕ looks like. Is 0.333 a good fit when $N = 150,000$ samples are used? We have plotted this ϕ in Figure 4. For ease of comparison with Figure 2 we have again presented only every tenth sample.

Content that we are on the right track we shall use p as the descriptor of our string. Our goal is now to devise a mathematical model that, at least, predicts behavior in line with the second and third columns of p.

3. A mathematical model

Implicit in all of the above is the assumption that the string's motion is planar and purely transverse to its rest state. Leaving this unchallenged (for the moment) we denote by ℓ the distance between the string's two supports and denote a material point by $(x, 0)$ where $x \in [0, \ell]$. At time t this material point lies at the point $(x, u(x,t))$. Assuming the string to be taut we expect zero displacement at its two ends, i.e.,

(3.1) $$u(0,t) = u(\ell, t) = 0, \qquad \forall t \geq 0.$$

In addition, we suppose that the pluck uniquely specifies both the position and velocity at each point in x at the initial instant of time. In symbols this means

(3.2) $$u(x,0) = u_0(x) \quad \text{and} \quad u_t(x,0) = v_0(x)$$

where u_0 and v_0 are known. Our task is to find and analyze the equation satisfied by $u(x,t)$ for $x \in (0, \ell)$ and all future time. Of the many possible paths to the string equation the most venerable is the one originating in the Calculus of Variations. The guiding principle goes variously under the names *Principle of Least Action, Principle of Stationary Action,* and *Hamilton's Principle.* Though already known to Leibniz and Euler the pronouncements of Maupertuis in 1746 in his "The laws of motion and of rest deduced from a metaphysical principle," brought it to life. In the words of Maupertuis,

"If there occurs some change in nature, the amount of action necessary for this change must be as small as possible."

In order to actually apply such a principle one must first quantify this mysterious 'action'. In the field of rigid-body mechanics, see, e.g., Arnold [A, p. 59] or Gelfand and Fomin [GF, §21], the action is the time integral of the difference of the kinetic and potential energies. The latter text, in §36, extends these notions to continua, such as our string, and argues that such bodies satisfy the Principle of **Stationary** (but not **Least**) Action. As the action of the string is the time average of the difference of kinetic and potential energies, the Principle of Stationary Action says that, on average, any variation in kinetic energy is balanced by a variation in the potential energy, and *vice versa*. Let us now express these energies and the associated action in terms of the string's displacement, $u(x,t)$.

The string's kinetic energy is half the product of its mass and the square of its velocity. The mass of the string is the product of its density, ρ, and length, ℓ. As $u(x,t)$ is the height of point x at time t, the velocity of the point is the time rate of change of u at x, i.e., $u_t(x,t)$. Summing over all points of the string we posit a kinetic energy of the form

$$T(u,t) \equiv \tfrac{1}{2}\rho \int_0^\ell u_t^2(x,t)\,dx.$$

Next, the potential energy is the work (product of force and distance) required to deform the string. The force in our context is the tension, τ, while the distance is the difference in lengths between the plucked and unplucked states. As the plucked state is assumed to be a graph, we arrive at a potential energy of the form

$$(3.3) \qquad U(u,t) = \tau \left\{ \int_0^\ell \sqrt{1 + u_x^2(x,t)}\,dx - \ell \right\}.$$

Left in place, the square root in U would clutter our application of the Principle of Stationary Action. Are there grounds for replacing the square root with something more agreeable? Even very strong plucks of the 60 cm string on the sonometer of Figure 1 produced maximal displacements of less than 1 cm. This suggests that $1/60$ is

Aye, There's the Rub

a reasonable upper bound for $|u_x(x,0)|$. As time increases the pluck travels along the string and then smoothes out and dissipates. That is, $|u_x(x,t)|$ is not likely to exceed $1/60$ for any $t \geq 0$. As a result, we may suppose that $u_x^2(x,t) < 1/3600$. As this is considerably smaller than 1, we may use $\sqrt{1+z} \approx 1 + z/2$ in (3.3) to arrive at

$$(3.4) \qquad U(u,t) = \tfrac{1}{2}\tau \int_0^\ell u_x^2(x,t)\,dx.$$

From T and U we now assemble the action

$$A(u) \equiv \int_{t_0}^{t_1} \{T(u,t) - U(u,t)\}\,dt,$$

over the time window $[t_0, t_1]$. The Principle of Least Action now asserts that the *actual* motion, u, is the one that minimizes A among all *possible* motions. The weaker statement that A is stationary at u will finally provide us with an equation for u. In order to make this precise let us compute the derivative A at u in the direction v,

$$A'(u) \cdot v \equiv \lim_{h \to 0} \frac{A(u + hv) - A(u)}{h}.$$

The direction v is chosen so that $u + hv$ is indeed a possible motion. By that we mean that $u + hv$ and u should satisfy the same boundary conditions with respect to both space and time. This means that

$$(3.5) \qquad v(0,t) = v(\ell, t) = 0 \quad \text{and} \quad v(x,t_0) = v(x,t_1) = 0.$$

Now

$$\begin{aligned}
A(u+hv) &- A(u) \\
&= \tfrac{1}{2}\int_{t_0}^{t_1}\int_0^\ell (\rho\{(u_t + hv_t)^2 - u_t^2\} - \tau\{(u_x + hv_x)^2 - u_x^2\})\,dx\,dt \\
&= h\int_{t_0}^{t_1}\int_0^\ell \{\rho u_t v_t - \tau u_x v_x + (h/2)(\rho v_t^2 - \tau v_x^2)\}\,dx\,dt,
\end{aligned}$$

and so, after dividing by h and letting h tend to zero, we find

$$(3.6) \qquad A'(u) \cdot v = \int_{t_0}^{t_1}\int_0^\ell (\rho u_t v_t - \tau u_x v_x)\,dx\,dt.$$

Now if u is indeed a stationary point of A, then (3.6) should vanish for every possible direction v. The consequences of this vanishing would be easier to read off if v rather than its derivatives were to appear in (3.6). To that end let us note that

$$u_t v_t = (u_t v)_t - u_{tt} v$$

and so, recalling (3.5),

$$\begin{aligned}\int_{t_0}^{t_1} u_t(x,t)v_t(x,t)\,dt &= \int_{t_0}^{t_1} ((u_t(x,t)v(x,t))_t - u_{tt}(x,t)v(x,t))\,dt \\ &= u_t(x,t)v(x,t)\Big|_{t=t_0}^{t=t_1} - \int_{t_0}^{t_1} u_{tt}(x,t)v(x,t)\,dt \\ &= -\int_{t_0}^{t_1} u_{tt}(x,t)v(x,t)\,dt.\end{aligned}$$

Similarly,

$$\int_0^\ell u_x(x,t)v_x(x,t)\,dx = -\int_0^\ell u_{xx}(x,t)v(x,t)\,dx.$$

On substitution of these last two eqautions into (3.6) we arrive at

(3.7) $\qquad A'(u) \cdot v = \int_{t_0}^{t_1}\int_0^\ell (\rho u_{tt}(x,t) - \tau u_{xx}(x,t))v(x,t)\,dx\,dt.$

If this indeed vanishes for every v on $(0,\ell) \times (t_0, t_1)$ satisfying (3.5), then necessarily the coefficient of v must vanish identically. This implication is typically called the *Fundamental Lemma of the Calculus of Variations*. Its application to (3.7) yields the so-called string equation,

(3.8) $\qquad \rho u_{tt}(x,t) - \tau u_{xx}(x,t) = 0 \quad \text{in} \quad (0,\ell) \times (t_0, t_1).$

This linear partial differential equation possesses an infinite number of independent solutions, namely

$$u(x,t) = f(x + t\sqrt{\tau/\rho}) + g(x - t\sqrt{\tau/\rho})$$

where f and g are *arbitrary* twice differentiable functions. As f and g describe *waves* with *speed* $\sqrt{\tau/\rho}$ one often speaks of (3.8) as the **wave equation**. One arrives at a unique solution to (3.8) by asking u to obey the boundary, (3.1), and initial, (3.2), conditions. In the absence of boundary conditions the reader may wish to check that it is a simple matter to express f and g in terms of u_0 and v_0. Boundaries produce reflections and therefore a more complicated representation for u. Our initial concern however is not with the *exact* values of u but rather with the question of how, or indeed whether, it decays.

One typically measures decay by studying the, so-called, instantaneous total energy

$$(3.9) \qquad E(t) \equiv T(u,t) + U(u,t) = \int_0^\ell (\rho u_t^2 + \tau u_x^2)\, dx.$$

Now by decay we mean that E should be decreasing. On evaluating its time derivative we find however that

$$E'(t) = 2\int_0^\ell (\rho u_{tt} u_t + \tau u_{xt} u_x)\, dx = 2\int_0^\ell (\rho u_{tt} - \tau u_{xx}) u_t\, dx = 0.$$

That is, the energy in the initial pluck is conserved throughout time. This of course condemns (3.8) as a model of what we observed in Figure 2. Recalling however that that signal was essentially a linear combination of damped sinusoids we might ask whether or not the string equation at least gets the frequencies right. This will require us to actually solve (3.8).

4. Solving the wave equation

By analogy to solving linear second-order ordinary differential equations we write (3.8) and (3.2) as the first-order system

$$(4.1) \qquad V_t = \mathcal{A} V, \qquad V(0) = \begin{pmatrix} u_0 \\ v_0 \end{pmatrix}$$

where

$$V \equiv \begin{pmatrix} u \\ u_t \end{pmatrix} \quad \text{and} \quad \mathcal{A} \equiv \begin{pmatrix} 0 & I \\ \frac{\tau}{\rho}\frac{d^2}{dx^2} & 0 \end{pmatrix}$$

and u is still assumed to satisfy the boundary conditions (3.1). This $\mathcal{A}$ is a matrix differential operator that acts on vectors composed of displacements and velocities. We define an inner (or dot) product for two such vectors, V and W. Namely, for two such vectors V and W we define

$$(4.2) \qquad \langle V, W \rangle \equiv \int_0^\ell \{\tau \partial_x V_1 \partial_x \overline{W}_1 + \rho V_2 \overline{W}_2\}\, dx,$$

where $\overline{W}$ denotes the complex conjugate of W. This definition is 'natural' in the sense that if $V = W = (u\ u_t)^T$, then $\langle V, V \rangle$ coincides with our earlier definition, see (3.9), of the instantaneous total energy of u. We shall denote by $\mathcal{E}$ the set of vector-valued functions V for which $\langle V, V \rangle < \infty$. (This 'finite energy' space has been well studied. In fact, it is the Hilbert Space $H_0^1(0, \ell) \times L^2(0, \ell)$, where the latter is the collection of square integrable functions on $(0, \ell)$ and the former is the collection of square integrable functions that vanish at 0 and ℓ and whose derivative is also square integrable.)

As in the ordinary matrix case one solves (4.1) in terms of the eigenvectors of $\mathcal{A}$, i.e., the solutions of

$$\mathcal{A}\Phi = \lambda \Phi.$$

It is not difficult to recognize these in the doubly infinite sequences

(4.3) $\qquad \Phi_{\pm n} = \sin(n\pi x/\ell) \begin{pmatrix} \frac{1}{n\pi}\sqrt{\frac{\ell}{\tau}} \\ \pm i \frac{1}{\sqrt{\ell\rho}} \end{pmatrix}, \quad \lambda_{\pm n} = \pm \frac{in\pi}{\ell}\sqrt{\frac{\tau}{\rho}}$

where $n = 1, 2, \ldots$. We have, for convenience, chosen $\Phi_{\pm n}$ to have unit length, i.e., $\langle \Phi_{\pm n}, \Phi_{\pm n} \rangle = 1$ in $\mathcal{E}$. On substituting the values of ρ, τ and ℓ associated with the string of §2 we find the eigenvalues

$$\lambda_{\pm n} = \pm in(701.83) = \pm i2\pi n(111.7).$$

These are in remarkable agreement (being close to integer multiples of 111) with the experimentally obtained resonant frequencies displayed in Figure 3. The fact that $\mathcal{A}$ has purely imaginary eigenvalues jibes with the fact that it is skew adjoint. Can you show that indeed $\langle \mathcal{A}V, W \rangle = -\langle V, \mathcal{A}W \rangle$?

A further consequence of skew adjointness is that the eigenvectors, $\{\Phi_{\pm n}\}_{n=1}^{\infty}$, of $\mathcal{A}$ constitute an orthonormal basis for the space $\mathcal{E}$. As a result, the full solution to (4.1) may be expressed as

(4.4) $\qquad V(t) = \sum_{n=1}^{\infty} \gamma_{\pm n} \exp(\lambda_{\pm n} t) \Phi_{\pm n}(x)$

where each $\gamma_{\pm n}$ encodes the projection of the initial pluck onto the respective eigenvector, i.e.,

$$\gamma_{\pm n} = \langle V(0), \Phi_{\pm n} \rangle = \sqrt{\frac{\tau}{\ell}} \int_0^\ell u_x(x,0) \cos(n\pi x/\ell)\, dx$$
$$\mp i \sqrt{\frac{\rho}{\ell}} \int_0^\ell u_t(x,0) \sin(n\pi x/\ell)\, dx.$$

Although the V of (4.4) is composed of terms that oscillate (at the right frequencies), none of them exhibits any decay.

5. The damped wave equation

Of the many ways that one may elicit decay, perhaps the simplest, by way of the analogy of adding a dashpot to a mass-spring system, is to introduce into the wave equation a term that is proportional to velocity, that is, to consider

(5.1) $$\rho u_{tt} - \tau u_{xx} + 2au_t = 0$$

for some constant a, in units of kg/m/s. Retaining our same notion of total energy we ask whether E now decreases when u satisfies (5.1) rather than (3.8). Note that its time derivative

$$E'(t) = 2 \int_0^\ell (\rho u_{tt} - \tau u_{xx}) u_t\, dx = -4a \int_0^\ell u_t^2\, dx$$

is negative so long as $a > 0$. That is, positive a produces decay of energy, E. So far so good, let us determine the effect of a on the eigenvalues and eigenvectors of the previous section.

With regard to the associated first-order system, (4.1), the $\mathcal{A}$ operator now takes the form

$$\mathcal{A}(a) = \begin{pmatrix} 0 & I \\ \frac{\tau}{\rho}\frac{d^2}{dx^2} & \frac{-2a}{\rho} \end{pmatrix}.$$

As a is constant the eigenvectors of $\mathcal{A}(a)$ are exactly as in (4.3). The eigenvalues become, however,

(5.2) $$\lambda_{\pm n} = -a/\rho \pm \sqrt{(a/\rho)^2 - (n\pi/\ell)^2 \tau/\rho}$$

and hence, so long as

$$a < \frac{\pi}{\ell}\sqrt{\frac{\tau}{\rho}}\rho \approx 1.057\frac{\text{kg}}{\text{m s}},$$

the real part of each eigenvalue is $-a/\rho$. The solution to (4.1) with $\mathcal{A}$ now replaced by $\mathcal{A}(a)$ remains (4.4) with the λ now given by (5.2). Although this indeed produces decay, it produces it in a far too uniform fashion. More precisely, each term decays at precisely the same rate, namely a/ρ. If we trust the larger variations in decay rates reported in the second column of (2.7), then we should search for a generalization of (5.1) that can exhibit such variable rates of decay. We now wish to argue that it suffices to let a vary with x.

When a varies with x, although expansions like (4.4) are still valid, we no longer have explicit expressions for the eigenfunctions and eigenvalues. For that reason we turn to their numerical approximation. More precisely, we solve

(5.3) $$\mathcal{A}(a)\Psi = \Lambda\Psi$$

by supposing Ψ to be a linear combination of the first $2m$ modes of the undamped problem. That is, we suppose

(5.4) $$\Psi = \sum_{k=1}^{m}(\gamma_k\Phi_k + \gamma_{-k}\Phi_{-k})$$

where the $\Phi_{\pm k}$ are specified in (4.3). On substituting (5.4) into (5.3) and taking the inner product of each side with one of these low undamped modes we arrive at

(5.5) $$\langle\mathcal{A}(a)\Psi,\Phi_j\rangle = \Lambda\langle\Psi,\Phi_j\rangle, \quad j = 1,\ldots,m,-1,\ldots,-m.$$

A moment's reflection now permits us to see in this the matrix eigenvalue problem

(5.6) $$G(a)\Gamma = \Lambda\Gamma$$

for

$$\Gamma = (\gamma_1\ \gamma_2\ \cdots\ \gamma_m\ \gamma_{-1}\ \gamma_{-2}\ \cdots\ \gamma_{-m})^T$$

where $G(a)$ is the $2m$-by-$2m$ matrix with elements

$$G_{j,k}(a) = \begin{cases} \langle \mathcal{A}(a)\Phi_k, \Phi_j \rangle & \text{if } 1 \le j \le m, \ 1 \le k \le m, \\ \langle \mathcal{A}(a)\Phi_k, \Phi_{m-j} \rangle & \text{if } m < j \le 2m, \ 1 \le k \le m, \\ \langle \mathcal{A}(a)\Phi_{m-k}, \Phi_j \rangle & \text{if } 1 \le j \le m, \ m < k \le 2m, \\ \langle \mathcal{A}(a)\Phi_{m-k}, \Phi_{m-j} \rangle & \text{if } m < j \le 2m, \ m < k \le 2m. \end{cases}$$

The inner product remains the one defined in (4.2). I hope that this notation does not obscure the relatively simple computations required to assemble $G(a)$. Perhaps we should write out in full what is going on in, say, the first line. If $1 \le j \le m$ and $1 \le k \le m$, then

$$\mathcal{A}(a)\Phi_k = \lambda_k \Phi_k + i\frac{1}{\rho\sqrt{\ell\rho}} \begin{pmatrix} 0 \\ -2a(x)\sin(k\pi x/\ell) \end{pmatrix}$$

and so

$$\langle \mathcal{A}(a)\Phi_k, \Phi_j \rangle = \lambda_k \langle \Phi_k, \Phi_j \rangle - \frac{2}{\ell\rho} \int_0^\ell a(x)\sin(k\pi x/\ell)\sin(j\pi x/\ell)\,dx$$

$$= \frac{ik\pi}{\ell}\sqrt{\frac{\tau}{\rho}}\delta_{j,k} - \frac{2}{\ell\rho} \int_0^\ell a(x)\sin(k\pi x/\ell)\sin(j\pi x/\ell)\,dx,$$

where $\delta_{j,k}$ is zero unless $j = k$, in which case it is one. So, at bottom, $G(a)$ is composed of integrals of the damping against products of sine functions. Upon constructing $G(a)$ we may use the `eig` routine in Matlab and arrive at the $2m$ matrix eigenvalues, $\{\Lambda_{\pm j}(a)\}_{j=1}^m$. As $G(a)$ inherits from $\mathcal{A}(a)$ the property of eigenvalues appearing in complex conjugate pairs we may, without loss, order them according to their imaginary parts, i.e.,

$$0 < \Im\Lambda_1(a) < \Im\Lambda_2(a) < \cdots < \Im\Lambda_m(a), \quad \Lambda_{-j} = \overline{\Lambda_j}.$$

Our goal now is to produce an a such that these matrix eigenvalues fit the second two columns of (2.7), namely the complex vector

(5.7) $$p_j^* \equiv p_{j,2} + i2\pi p_{j,3}.$$

In the interest of practicality we limit our search to a finite-dimensional class of dampings. In keeping with the above choice of basis we suppose that

$$a(x) = \sum_{k=1}^n \alpha_k \sin((2k-1)\pi x/\ell).$$

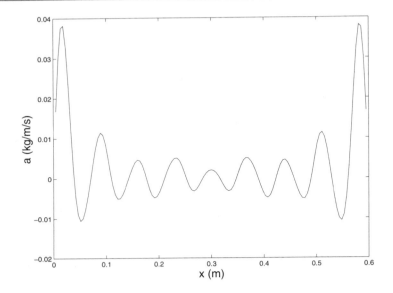

Figure 5. The rub

Our limitation to odd sine terms reflects our belief that the 'proper' a ought to be symmetric with respect to the midpoint, $x = \ell/2$. Finally, to fit the $\Lambda_j(a)$ to the p_j^* is to solve

$$(5.8) \qquad \min_{\alpha \in \mathbf{R}^n} \sum_{j=1}^{m} |p_j^* - \Lambda_j(a)|^2.$$

Of course, as the p_j^* constitute only 16 real parameters we may only hope to find a unique minimizer when $n \leq 16$. Asking for the most, we take $n = 16$ and solve (5.8) in Matlab via `leastsq`. We have plotted the resulting a in the figure below.

This result indeed exposes the sought after rub (see Figure 5). Namely, a is large at the two ends, signifying that through rubbing against its supports the string eventually sheds the energy delivered by the pluck.

Of course one should not accept the veracity of a model based on the result of a single experiment. Rather than simply replucking the same string, however, we wish to see whether our methods might be able to distinguish the presence of an artificial damper.

6. Discerning the presence of additional damping

It is not unnatural to attempt to identify the magnitude and position of dissipative mechanisms. As a simple example, engineers seek means by which the size and location of a leak in a cable may be discerned from measurements taken near the cable's ends. We therefore add a dissipative mechanism, a pair of attracting magnets at the center of the string (recall Figure 1), and attempt to detect it by the eigenvalue matching method used above.

On placing the detector 15 cm to the right of the magnetic damper, we pluck the string and record the data plotted in Figure 6 (top).

Comparing this to Figure 2 we note that the magnetic damper has indeed increased, or accelerated, the decay. To ascertain the associated natural frequencies we again turn, see Figure 6 (bottom), to the Discrete Fourier Transform. Comparing the latter with Figure 3 we note that, although the resonant frequencies are essentially unchanged, the peak associated with the lowest frequency is severely diminished. That the lowest frequency is indeed the one that suffers the greatest dissipation is borne out by fitting the response of Figure 6 (top) to the sum of damped sinusoids referred to as ϕ; recall (2.5). The p that solves (2.6) in this case is

$$p = \begin{pmatrix} -0.0059 & -1.3034 & 107.3376 & -1.6096 \\ -0.0228 & -0.2567 & 224.1213 & -1.4830 \\ -0.0048 & -0.4860 & 335.2361 & -1.0603 \\ -0.0027 & -0.2835 & 448.5797 & -2.6720 \\ 0.0012 & -0.5176 & 560.4651 & -5.5950 \\ -0.0074 & -0.6233 & 673.8908 & -7.3395 \\ -0.0017 & -0.7285 & 786.0627 & -10.1866 \\ -0.0008 & -0.9328 & 898.1001 & -10.7800 \end{pmatrix}.$$

Notice that the decay, $p(1,2)$, of the lowest frequency, is 2 to 5 times greater than that of the next few. From this p we build the p^* in

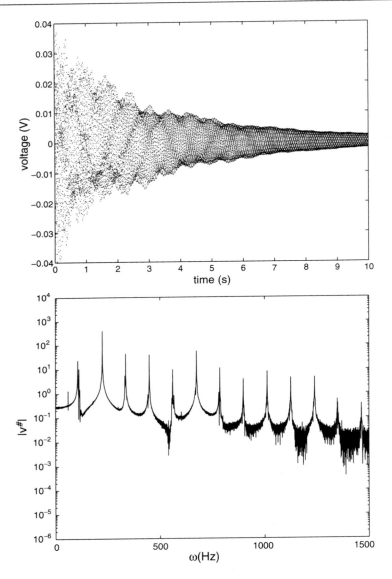

Figure 6. The pluck response with damper and the magnitude of its DFT

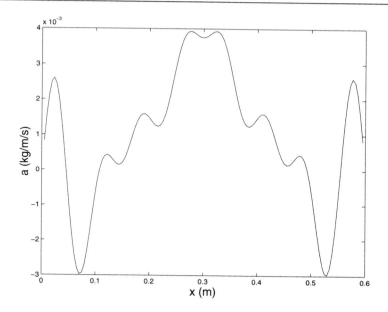

Figure 7. The magnetic rub

(5.7) and solve (5.8). Finding little improvement in the fit for $n > 8$ we plot in Figure 7 the best fit for $n = 8$.

Even with its ups and downs, this result is a strong indication that a dissipative device has been placed near the midpoint of the string.

7. Concluding remarks

In my attempt at flushing out the rub I have invoked techniques and results from a number of fields. In the interest of both always keeping the goal in sight and actually producing an 'answer' you may accuse me of having taken giant leaps or, perhaps worse, having pursued my quarry without regard to rigor. Given the additional constraints of space, I can only answer these criticisms by directing you to the relevant literature.

Regarding §2, even after more than 100 years, Rayleigh's text remains modern. It serves as a fascinating (and inexpensive) supplement to an introduction to partial differential equations. The engineering field of identifying dynamical systems by their modes of vibration is called Modal Analysis. Along these lines I highly recommend the text of Ewins [E].

For more on the history and application of the Principle of Least Action the best two sources are the lovely text of Hildebrandt and Tromba [HT] and the lecture of Feynman [FLS]. For an alternate derivation of the wave equation I recommend the text by Knobel [K].

For a more careful derivation of the eigenvalues and eigenvectors of the undamped and damped wave operators I recommend the text of Weinberger [W]. You will also find there a study of strings that may undergo *longitudinal* as well as transverse motion. For those with a solid grounding in analysis, the text of Treves [T] establishes the properties of the finite energy space, $\mathcal{E}$, and the operator, $\mathcal{A}$, necessary for a rigorous interpretation of (4.4).

To learn more about the approximation of operator eigenvalues by matrix eigenvalues, as practiced in §5, I would turn to the text of Chatelin [Ch].

Finally, you may ask, do the eigenvalues of the damped wave operator, $\mathcal{A}(a)$, indeed uniquely determine the damping a? We have offered here no more than numerical evidence in support of the conjecture. The proof, see Cox and Knobel [CK], relies on the clever implementation by Yamamoto of the so-called Gelfand–Levitan transform. For the role of this transform in resolving questions of the type posed here I recommend the text of Levitan [L].

Bibliography

[A] V.I. Arnold, *Mathematical Methods of Classical Mechanics*, Springer-Verlag, 1989.

[Ch] F. Chatelin, *Spectral Approximation of Linear Operators*, Academic Press, 1983.

[CK] S.J. Cox and R. Knobel, *An inverse spectral problem for a nonnormal first order differential operator*, Integral Equations and Operator Theory, 25 (1996), pp. 147–162.

[E]   D.J. Ewins, *Modal testing : theory and practice*, Wiley, 1984.
[FLS] R.P. Feynman, R.B. Leighton and M. Sands, *The Feynman Lectures on Physics*, Addison-Wesley, 1963.
[FR] N.H. Fletcher and T.D. Rossing, *The Physics of Musical Instruments*, Springer-Verlag, 1991.
[GF] I.M. Gelfand and S.V. Fomin, *Calculus of Variations*, Prentice-Hall, 1963.
[HT] S. Hildebrandt and A. Tromba, *The Parsimonious Universe : Shape and Form in the Natural World*, Copernicus, 1996.
[K]   R. Knobel, *An Introduction to the Mathematical Theory of Waves*, Student Mathematical Library, Vol. 3, AMS, 2000.
[L]   B.M. Levitan, *Inverse Sturm-Liouville Problems*, VSP, 1987.
[R]   J.W.S. Rayleigh, *The Theory of Sound*, Dover, 1945.
[T]   F. Treves, *Basic Linear Partial Differential Equations*, Academic Press, 1975.
[W]   H.F. Weinberger, *A First Course in Partial Differential Equations with Complex Variables and Transform Methods*, New York, 1965.

Proof of the Double Bubble Conjecture

Frank Morgan

1. The news. In 2000, Hutchings, Morgan, Ritoré and Ros ([11]; see [4], [13]) announced a proof of the Double Bubble Conjecture, which says that the familiar double soap bubble of Figure 1 provides the least-area way to enclose and separate two given volumes of air. The two spherical caps are separated by a third spherical cap, all meeting at 120 degree angles; if the volumes are equal, the separating surface is a flat disc. This result is the culmination of ten years of remarkable progress by many mathematicians, including undergraduates.

2. The planar double bubble. It all started when the 1990 Williams College NSF "SMALL" undergraduate research Geometry Group [6] proved the Planar Double Bubble Theorem: the standard double bubble of Figure 2a, b provides the least perimeter way to enclose and separate two regions of prescribed area in the plane. Perimeter counts every piece of curve once, whether it is on the exterior or the interior.

This article is reprinted from Amer. Math. Monthly **108** (2001), 193–205 with permission.

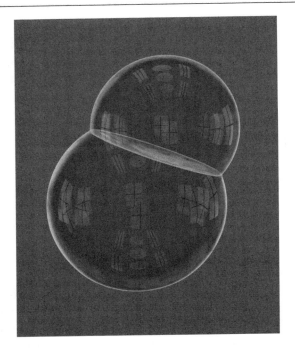

Figure 1. The standard double bubble provides the least-perimeter way to enclose and separate two prescribed volumes. Copyright John M. Sullivan, University of Illinois; color version at www.math.uiuc.edu/-jms/Images.

The group leader, Joel Foisy, in his subsequent undergraduate thesis, apparently gave the first statement of the Double Bubble Conjecture in $\mathbf{R}^3$ as a conjecture. Plateau [14, pp. 300-301] had studied the double bubble over a hundred years earlier, and Boys [1, p. 120] had spoken of the conjecture as a fact, as it was widely accepted for many years.

3. Proof of the Planar Double Bubble Theorem.

The major difficulty in the proof is showing that each region and the exterior is connected. Although it may seem obvious that having several components scattered throughout the bubble as in Figure 2c, d, e could not minimize the perimeter, this turns out to be not so easy to prove. Of course we prefer to *prove*, rather than to *assume*, that the regions

The Double Bubble Conjecture

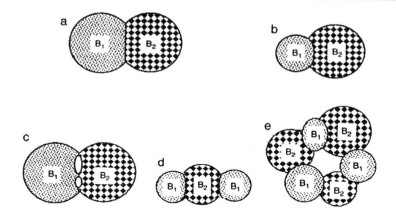

Figure 2. The standard planar double bubble (a and b), and not some exotic alternative with disconnected regions or empty chambers (c, d, or a), provides the least-perimeter way to enclose and separate two regions of prescribed area, as proved by the 1990 undergraduate research Geometry Group [6, Figs. 1.0.1, 1.0.2].

are connected. Besides, in higher dimensions a minimizer with disconnected regions could arise as a limit of regions connected by thin tubes as the tubes shrink away.

The original ingenious proof has undergone several simplifications, largely due to Michael Hutchings. The simplest argument starts with any competitor and deforms it to a standard double bubble while decreasing the perimeter. Unlike the original proof, it does not need to assume that a nice minimizer exists; it does not depend on the general existence and regularity theory.

During the argument it is convenient to work in the category of the "overlapping" bubbles of the 1992 Geometry Group [2, 3], smooth *immersions* of finite embedded planar graphs, in which the faces may overlap as in Figure 3.

In Section 4 we show that for prescribed areas there is a unique standard double bubble. consisting of three circular arcs meeting at 120 degrees; if the areas are equal, the separating curve is a straight line segment.

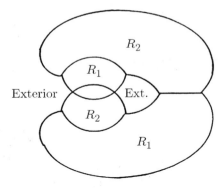

Figure 3. In this "overlapping" bubble, components of the regions R_1 and R_2 overlap. Copyright 2000 Frank Morgan.

Figure 4. Vertices of degree four or more may be reduced. Copyright 2000 Frank Morgan.

Given areas A_1 and A_2, consider any, perhaps overlapping, double bubble with areas *at least* A_1 and A_2. We may assume there is no such bubble of fewer faces with smaller perimeter. We may always assume that all vertices have degree at least three.

First we claim that there are no empty chambers (bounded components of the exterior). Otherwise we could delete an edge, incorporate the empty chamber into the neighboring region, and reduce the number of faces.

Second we claim that there are no vertices of degree greater than three. Otherwise two of the faces at the vertex both belong to the first region, to the second region, or to the exterior. You can always combine these faces while reducing the perimeter and maintaining areas, as suggested in Figure 4.

The Double Bubble Conjecture

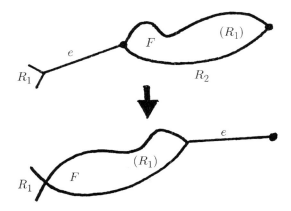

Figure 5. Flipping a face of two edges and an adjacent edge creates a vertex of degree four. Copyright 2000 Frank Morgan.

Consider the dual graph formed by placing a vertex inside each face of the two regions, with an edge between vertices of adjacent faces. Because the exterior is connected, this dual graph has no cycles. Hence there is some face F that lies at an endpoint of the dual graph. This face F must have exactly two edges and exactly two vertices, as in Figure 5.

Unless the bubble is already combinatorially the standard bubble, take out F and one neighboring edge e, flip it over end to end, and reinsert it as in Figure 5. This operation produces a possibly overlapping double bubble with a vertex v of degree four, a contradiction. We conclude that the bubble is combinatorially the standard bubble, consisting of three edges meeting at two points.

Replace the three edges by circular arcs, maintaining areas. By the isoperimetric property of circles and circular arcs, any such replacement reduces perimeter. Now we may assume that the bubble is embedded.

Next minimize the perimeter in the category of three circular arcs enclosing given areas. We claim that the edges must meet at 120-degree angles. Otherwise you could deform the bubble a bit so as to preserve areas and decrease length. After this deformation the edges would not be arcs of circles, but you could replace them with circular arcs, so as to decrease the length even more.

Now we have a standard bubble of areas at least A_1 and A_2. Since by reducing either area you can reduce the perimeter, the standard bubble of areas exactly A_1 and A_2 has even smaller perimeter.

Thus starting with any other competitor, we have reduced the perimeter and arrived at the standard double bubble, which must therefore be perimeter minimizing.

Before going further, we verify that there is a unique standard double bubble in any $\mathbf{R}^n$ $(n \geq 2)$.

4. The standard double bubble. *For prescribed volumes v, w, there is a unique standard double bubble in $\mathbf{R}^n$ consisting of three spherical caps meeting at 120 degrees as in Figure 1.*

Proof. Consider a unit sphere through the origin and a congruent or smaller sphere intersecting it at the origin (and elsewhere) at 120 degrees as in Figure 6. There is a unique completion to a standard double bubble. Varying the size of the smaller sphere yields all volume ratios precisely once. Scaling yields all pairs of volumes precisely once.

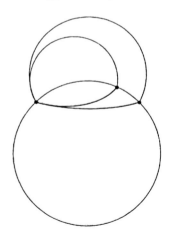

Figure 6. Varying the size of the smaller component yields all volume ratios precisely once. Copyright 2000 Frank Morgan.

The Double Bubble Conjecture

5. Existence and regularity. Proving the existence of a perimeter-minimizing double bubble of prescribed volumes in $\mathbf{R}^n$ is no easy matter. Classical spaces of surfaces are not compact, especially if there is no *a priori* bound on topological complexity or how the pieces fit together. Furthermore, one cannot consider only bubbles with connected regions, because they might in principle disconnect in the minimizing limit as thin connecting tubes shrink away.

Geometric measure theory (see [13]), as developed in the 1950s and 1960s by L. C. Young, E. De Giorgi, H. Federer, and W. Fleming, provides very general compact spaces of surfaces of bounded diameter in $\mathbf{R}^n$. Soap bubbles provide additional difficulties, because there is no easy *a priori* bound on the diameter. In 1976, F. Almgren provided a general proof using geometric measure theory of the existence of perimeter-minimizing soap bubble clusters, and with J. Taylor proved that in $\mathbf{R}^3$ they consist of smooth constant-mean-curvature surfaces meeting in threes at 120 degrees along curves, which in turn may meet in fours at equal angles of about 109 degrees ($\cos^{-1}(-1/3)$).

The following key symmetry theorem is based on an idea of Brian White, written up by Foisy [5, Thm. 3.4] and Hutchings [12, Thm. 2.6]. Since it reduces the Double Bubble Conjecture to questions about curves in the plane, it was the main reason for mistakenly considering the matter settled.

6. Symmetry Theorem. *A perimeter-minimizing double bubble B in $\mathbf{R}^n$ is a surface of revolution about a line.*

Proof sketch, case $n = 3$. First we claim that there are two orthogonal planes that split both volumes in half. Certainly, for every $0 \leq \theta \leq \pi$, there is a vertical plane at angle θ to the xz-plane that splits the first region in half. These planes can be chosen to vary continuously back to the original position, now with the larger part of the second region on the other side. Hence for some intermediate θ, the plane splits both volumes in half. Turning everything to make this plane horizontal and repeating the argument yield a second plane, as desired. Hence we may assume that the xz- and yz-planes bisect both volumes.

Actually, we need to modify the process a bit. After obtaining the first plane, reflecting the half of B of least area yields a bubble of no more area. It must therefore have the same area, and either half would work. If the original B were not a surface of revolution, neither is some such half plus reflection. Hence we may assume that B is symmetric under reflection across each plane, and hence under their composition, rotation by 180 degrees about the z-axis. Hence every plane containing the z-axis splits both volumes in half.

We claim that at every regular point, the bubble B is orthogonal to the vertical plane. Otherwise the smaller or equal half of B, together with its reflection, would be a minimizer with an illegal singularity, which could be smoothed to reduce area while maintaining volume. Now it follows that B is a surface of revolution.

Hutchings realized that the symmetry proof could be generalized to prove the following monotonicity result, which is not as obvious as it sounds.

7. Monotonicity [12, THM. 3.2]. *The least area $A(v,w)$ of a double bubble of volumes v, w in $\mathbf{R}^n$ is a nondecreasing function in each variable.*

Proof sketch, case $n = 3$. We prove that for fixed w_0, $A(v, w_0)$ is nondecreasing in v. If not, then there is a local minimum at some v_0. For simplicity, we treat just the case of a strict local minimum. Consider a minimizing double bubble B of volumes v_0, w_0. By the Symmetry Theorem 6 and its proof, B is a surface of revolution about a line $L = P_1 \cap P_2$, where P_1 and P_2 are planes that divide both regions in half. Choose a plane P_3 near P_2 that divides the second region in half but does not contain L. We claim that it divides the first region in half. Otherwise, the half with smaller or equal area, reflected across the plane, would yield a bubble of no more area and slightly different volume, contradicting the assumption that v_0 is a strict local minimum. Therefore P_3 splits both volumes in half. Now as in the proof of Symmetry 6, B is symmetric about the line $L' = P_1 \cap P_3$ as well as about L. It follows that B consists of concentric spheres, which is impossible. Therefore $A(v, w_0)$ must be nondecreasing as desired.

The Double Bubble Conjecture

8. Corollary (Connected Exterior) [12, THM. 3.4]. *An area-minimizing double bubble in $\mathbf{R}^n$ has connected exterior ("no empty chambers").*

Proof. If the exterior has a second, bounded, component, removing a surface to make it part of one of the two regions would reduce the area and increase the volume, in contradiction to Monotonicity 7.

Symmetry and Connected Exterior are the primary lemmas for the following structure theorem.

9. Hutchings Structure Theorem [12, THM. 5.1]. *An area-minimizing double bubble in $\mathbf{R}^n$ is either the standard double bubble or another surface of revolution about some line, consisting of two round spherical caps with a toroidal innertube, successively layered with more toroidal innertubes, as in Figure 7. The surfaces are all constant-mean-curvature surfaces of revolution, "Delaunay surfaces", meeting in threes at 120 degrees.*

Figure 7. A nonstandard area-minimizing double bubble in $\mathbf{R}^n$ consists of a central bubble with layers of toroidal bands. Copyright 2000 Yuan Y. Lai.

Comments on the proof. Once the exterior is known to be connected, there are not many possibilities for a double bubble of revolution. A contraction argument given by Foisy [5, Thm. 3.6] shows that the bubble must intersect the axis.

There can be at most two spherical caps and at most one toroidal innertube directly on the spherical caps. Indeed, if there were a band of a third sphere S between two toroidal innertubes in the first layer, the rest of the bubble would consist of two big pieces, which could be rolled around S to touch each other and create an illegal singularity.

Combinatorial finiteness follows from the following bound.

10. Hutchings component bound. *Consider a minimizing double bubble of volumes v, $1-v$ in $\mathbf{R}^n$. Then the number k of components of the first region satisfies*

(1) $$A(v)k^{1/n} \leq 2A(v, 1-v) - A(1) - A(1-v).$$

Here $A(v)$ is the surface area of a single bubble (round sphere) of volume v, and $A(v, 1-v)$ is the (unknown) area of a minimizing double bubble, which fortunately is bounded above by the area of the standard double bubble.

Proof sketch. The decomposition suggested by Figure 8 gives a lower bound on $A(v, 1-v)$ when the first region has a small component of volume x. Furthermore, Hutchings generalizes Monotonicity 7 to show that $A(v,w)$ is strictly concave in both variables. Now the desired component bound follows by just a little algebra.

Remark. Mathematica graphs of the bound (1) on the number of components k for $\mathbf{R}^2$, $\mathbf{R}^3$, $\mathbf{R}^4$, and $\mathbf{R}^5$, with $A(v, 1-v)$ replaced by the larger or equal area of the standard double bubble, appear in Figure 9. Some results are summarized in Table 10. In particular, in $\mathbf{R}^2$ both regions are connected, from which the Double Bubble Conjecture in $\mathbf{R}^2$ follows easily. The $\mathbf{R}^n$ bounds are elegantly deduced from the Hutchings Bound 10 by the 1999 Geometry Group [10, Prop. 5.3].

The following conjecture [10, Conj. 4.10] would provide an elegant way to prove the numerical bounds rigorously, a task otherwise quite awkward.

The Double Bubble Conjecture

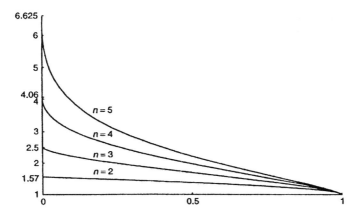

Figure 8. This decomposition gives a lower bound on $A(v, 1-v)$ when the first region has a small component of volume x. Each piece of surface on the left occurs exactly twice on the right. Drawing by James F. Bredt, copyright 2000 Frank Morgan.

Figure 9. A bound on the number of components of the first region in a minimizing double bubble of volumes v, $1 - v$ in $\mathbf{R}^2$ through $\mathbf{R}^5$. Figure by Ben W. Reichardt [10, Fig. 2].

	$\mathbf{R}^2$	$\mathbf{R}^3$	$\mathbf{R}^4$	$\mathbf{R}^5$	$\mathbf{R}^n$
Bounds on number of components in larger or equal region	1	1	1	2	3
Bounds on number of components in smaller region	1	2	4	6	2^n

Table 10. Bounds on the number of components in a perimeter-minimizing double bubble.

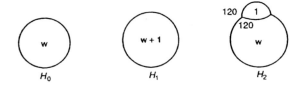

Figure 11. Conjecture: *The pictured curvatures satisfy $H_2 > (H_0+H_1)/2$.* If true, this conjecture would provide an elegant way to prove the bounds of Table 10. Copyright 2000 Frank Morgan.

11. Conjecture.[*] *In $\mathbf{R}^n$, let H_0, H_1, H_2, respectively, denote the mean curvature of a sphere of volume w, a sphere of volume $w+1$, and the exterior of the second region of the standard double bubble of volumes 1, w as suggested by Figure 11. Then*

$$2H_2 > H_0 + H_1.$$

Note that the Hutchings Bound implies that for equal volumes, each region of a minimizing double bubble in $\mathbf{R}^3$ is connected. In 1995 Hass and Schlafly exploited this information to prove this case of the Double Bubble Conjecture *by computer*.

12. Theorem ([7], [9]). *For equal volumes in $\mathbf{R}^3$ the standard double bubble uniquely minimizes the perimeter.*

Proof sketch ([9] and [8]). For equal volumes in $\mathbf{R}^3$, the Hutchings theory says that both regions are connected (Table 10) and that a minimizer consists of a central bubble with a toroidal innertube around it as in Figure 12. The innertube need not be centered around the waist, but there is at most a 2-parameter family of possibilities. Hass and Schlafly used a computer to recursively divide the parameter domain into cases and subcases, always seeking some contradiction. Often the volumes fail to be equal. Sometimes there are unstable pieces of surface, as in Figure 13. Sometimes the pieces just do not fit together. In all computations, the values are bounded above and below by exact computations in integer arithmetic. If no contradiction appears with the required accuracy, the case is subdivided into finer subcases. If

---

[*]Marilyn Daily apparently recently has proved Conjecture 11.

The Double Bubble Conjecture

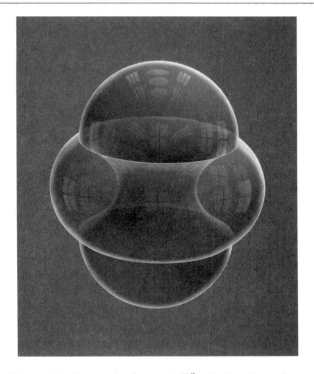

Figure 12. For equal volumes in $\mathbf{R}^3$, the Hutchings theory says that a perimeter-minimizing double bubble consists of a central bubble with a single toroidal innertube around it. Copyright John M. Sullivan. Color version at www.math.uiuc. edu/~jms/Images.

eventually every case terminates in a contradiction and the computer program halts, the theorem is proved. In fact, in 1995 it finished on a PC in about twenty minutes, having computed 200,260 numerical integrals.

For arbitrary volumes in $\mathbf{R}^3$, the Hutchings Bound (Table 10) says only that a minimizing double bubble has at most three components, as in Figure 14, much too complicated a family of possibilities to be handled by current computer technology. The computer-free 2000 proof uses a new instability argument.

Figure 13. Candidates with large oscillations may be eliminated as unstable. Copyright Joel Hass, Jim Hoffman, and Roger Schlafly. Color version available at www.math.ucdavis.edu/~hass/bubbles.html.

13. Theorem [11]. *For given volumes in $\mathbf{R}^3$, the standard double bubble of Figure 1 uniquely minimizes the perimeter.*

Proof sketch. The idea is to show that any nonstandard candidate is unstable; that is, it can be deformed so as to reduce the area while preserving both volumes. The desired instability is obtained by rotating different pieces of the double bubble, such as the left and right halves of Figure 15, in different directions or at different rates ω_i around a carefully chosen axis A.

In general, maintaining the volume constraints requires two extra degrees of freedom, requiring four rather than two pieces. To avoid

The Double Bubble Conjecture

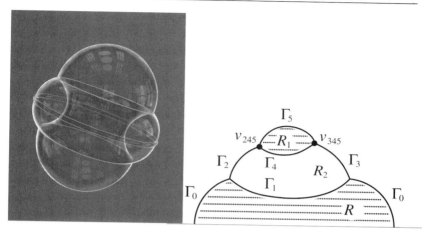

Figure 14. A minimizing double bubble in $\mathbf{R}^3$ has at most three components. Copyright John M. Sullivan, University of Illinois (color version at www.math.uiuc.edu/~jms/Images); second image from [11].

ripping the bubble apart, the bubble must be tangential to the rotational vector field where the pieces meet. Figure 15 could be divided into the four quadrants.

Of course, if every $\omega_i = 1$, the whole bubble just rotates, and the second variation of perimeter vanishes. If, as we assume to obtain a contradiction, the bubble is stable, so that no choice of ω_i's yields a negative second variation, it follows that every choice of ω_i yields a vanishing second variation, corresponding to the solution to some partial differential equation. If there are (at least) four pieces, then the rates of rotation ω_1, ω_2, ω_3, ω_4 can be chosen to maintain the two volumes, with at least one but not all of the ω_i's equal to 0. Since one ω_i vanishes, by unique continuation for solutions to nice partial differential equations, all the variations must vanish and the relevant pieces of the bubble must be spherical, which leads to a contradiction.

Can you always find an axis A of rotation so that the curves where the bubble is tangential divide the bubble into four pieces?

We carefully choose the axis A perpendicular to the axis of rotational symmetry, the x-axis. The surface is automatically tangential

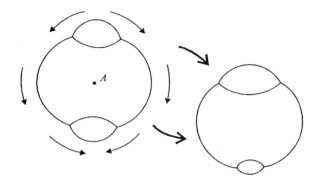

Figure 15. Rotating the two halves of this bubble in opposite directions about a new axis A stretches the top, shrinks the bottom, and reduces the area. Copyright 2000 Frank Morgan.

in the plane normal to A, so we always have at least two pieces, top and bottom.

If as in the case of equal volumes the bubble always had just two components as in Figure 16, a suitable axis A is provided by the perpendicular bisector of the two vertices V_1, V_2: at the closest and the most distant intermediate points, p_1 and p_2, the radius vector from A meets the bubble orthogonally, so that the bubble is tangential at each p_i and at the whole circular orbit of p_i around the x-axis. The bubble divides into the four requisite pieces. This proves the Double Bubble Conjecture for the case of equal volumes.

For the general case of three components, the Euclidean geometry is much more complicated but manageable. One considers cases according to the relative position of parts of the bubble, as in Figure 17.

14. Higher dimensions. In an amazing postscript, the 1999 Williams College "SMALL" Geometry Group, consisting of Ben Reichardt, Cory Heilmann, Yuan Lai, and Anita Spielman, has extended the proof of the Double Bubble Conjecture to $\mathbf{R}^4$ and certain higher dimensional cases.

The Double Bubble Conjecture

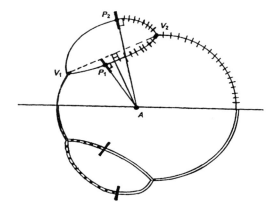

Figure 16. If the bubble has just two components, a central bubble and toroidal innertube, a suitable axis of instability A is provided by the perpendicular bisector of the two vertices V_1, V_2. The places " — " where the rotational vector field is tangent to the surface divide the bubble into four pieces. Copyright 2000 Frank Morgan.

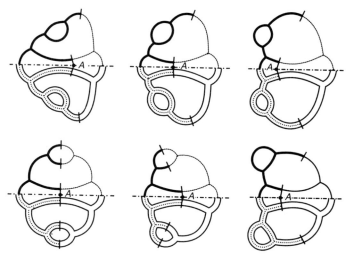

Figure 17. If the bubble has three components, one has to find the axis A of instability case by case, according to the relative position of parts of the bubble. Drawing by James F. Bredt, copyright 2000 Frank Morgan.

15. Theorem [15, THMS. 9.1, 9.2]. *In $\mathbf{R}^4$, the standard double bubble is the unique minimizer. In $\mathbf{R}^n$, for prescribed volumes $v > 2w$, the standard double bubble is the unique minimizer.*

Proof sketch. In these cases, the Hutchings theory implies that the larger region is connected. Since there is no bound on the number of components of the second region, the case-by-case analysis of Figure 16 must be broken down into more general arguments about constituent parts.

16. Open questions. Hutchings et al. [11, Intro.] conjecture that the standard double bubble in $\mathbf{R}^n$ is the unique stable double bubble. Although the conclusion of their proof shows the final competitor unstable, earlier portions such as Symmetry 6 assume area minimization, so that there could be unsymmetric stable bubbles, for all we know.

Sullivan [16, Prob. 2] has conjectured that the standard k-bubble in $\mathbf{R}^n$ ($k \leq n+1$) is the unique minimizer enclosing k regions of prescribed volume. There is also the very physical question in $\mathbf{R}^3$ of whether the standard double bubble is the unique stable double bubble with connected regions; by [11, Cor. 5.3], it would suffice to prove rotational symmetry.

Acknowledgments. This article is largely based on the new Chapter 14 of the 2000 edition of [13], where further details and references may be found. This work was partially supported by a National Science Foundation grant.

Bibliography

[1] C. V. Boys, *Soap-Bubbles*, Dover, New York, 1959.

[2] Christopher Cox, Lisa Harrison, Michael Hutchings, Susan Kim, Janette Light, Andrew Mauer, and Meg Tilton, The shortest enclosure of three connected areas in $\mathbf{R}^2$, *Real Anal. Exchange* 20 (1994/95) 313-335.

[3] Christopher Cox, Lisa Harrison, Michael Hutchings, Susan Kim, Janette Light, Andrew Mauer, and Meg Tilton, The standard triple

bubble type is the least-perimeter way to enclose three connected areas, NSF "SMALL" undergraduate research Geometry Group report, Williams College, 1992.

[4] Barry Cipra, Why double bubbles form the way they do, *Science* 287 (17 March 2000) 1910-1911.

[5] Joel Foisy, Soap bubble clusters in $\mathbf{R}^2$ and $\mathbf{R}^3$, undergraduate thesis, Williams College, 1991.

[6] Joel Foisy, Manuel Alfaro, Jeffrey Brock, Nickelous Hodges, and Jason Zimba, The standard double soap bubble in $\mathbf{R}^2$ uniquely minimizes perimeter, *Pacific J. Math.* 159 (1993) 47-59.

[7] Joel Hass, Michael Hutchings, and Roger Schlafly, The double bubble conjecture, *Electron. Res. Announc. Amer. Math. Soc.* 1 (1995) 98-102.

[8] Joel Hass and Roger Schlafly, Bubbles and double bubbles, *Amer. Scientist*, Sept-Oct, 1996, 462-467.

[9] Joel Hass and Roger Schlafly, Double bubbles minimize, *Ann. Math.* 151 (2000) 459-515.

[10] Cory Heilmann, Yuan Y. Lai. Ben W. Reichardt, and Anita Spielman, Component bounds for area-minimizing double bubbles, NSF "SMALL" undergraduate research Geometry Group report, Williams College, 1999.

[11] Michael Hutchings, Frank Morgan, Manuel Ritoré, and Antonio Ros, Proof of the double bubble conjecture, *Electron. Res. Announc. Amer. Math. Soc.* 6 (2000) 45-49. Full paper now published as: Michael Hutchings, Frank Morgan, Manuel Ritoré, and Antonio Ros, Proof of the double bubble conjecture. *Ann. of Math. (2)* 155 (2002), no. 2, 459–489.

[12] Michael Hutchings, The structure of area-minimizing double bubbles, *J. Geom. Anal.* 7 (1997) 285-304.

[13] Frank Morgan, *Geometric Measure Theory: a Beginner's Guide*, third ed., Academic Press, Boston, 2000.

[14] J. Plateau, *Statique Expérimentale et Théorique des Liquides Soumis aux Seules Forces Moléculaires*, Paris, Gauthier-Villars, 1873.

[15] Ben W. Reichardt, Cory Heilmann, Yuan Y. Lai, and Anita Spielman, Proof of the double bubble conjecture in $\mathbf{R}^4$ and certain higher dimensions, *Pacific J. Math.* 208 (2003), no. 2, 347–366.

[16] John M. Sullivan and Frank Morgan, eds., Open problems in soap bubble geometry, *Internat. J. Math.* 7 (1996) 833-842.

Minimal Surfaces, Flat Cone Spheres and Moduli Spaces of Staircases

Michael Wolf

My formal goal in this chapter is to show you the proof of the existence of some new minimal surfaces discovered by my collaborator, Matthias Weber of Universität Bonn, and myself in 1996. I say "formal goal" because my secret principal goal is to wander through several areas of classical mathematics and several concepts in contemporary mathematics and show you how smoothly they fit together.

I should explain this perspective a bit more carefully before we really set out (as it is a bit different from a typical introduction to a subject). To really understand all of the details of the proof, you'd need a pretty substantial background in geometry and complex analysis, much more than we are assuming in this book. On the other hand, the outline of the proof is not only accessible to readers who are in the middle of their undergraduate careers, but I think it also

---

©2004 by the author

serves as an introduction to modern mathematical study. One of the most enchanting aspects of mathematics research is how the study of a single problem will force you to travel from one field of mathematics to another and another. How, then, can you possibly ever prepare for the study of a problem? It seems as though you'd have to have taken a serious course in most aspects of mathematics, and you'd spend all of your life preparing and none of your life actually participating. The answer is that most of the time you learn what you need — and maybe a little bit more — as you go along, and that is how I'm structuring this chapter: we will describe the context and statement of one result and then develop the architecture of its proof as we follow a meandering path through several fields of mathematics in which you may or may not have a complete background — for the purposes here, it doesn't matter so much if you don't. I hope we also encounter some ideas and concepts that are important in present-day mathematics — these are ideas that are often not met in undergraduate and beginning graduate classes. So my caveat is that I do not plan a balanced and comprehensive introduction to minimal surfaces; see [**Oss86**], [**DHKW92**] for that. Instead, I plan a path through interesting terrain towards a site in minimal surface theory of interest to me.

1. Minimal surfaces

Let's begin with the notion of a minimal surface. It's natural to begin here, as this is a topic informed by physical experiments on soap films, most notably those by the (then blind) 19th century Belgian physicist Plateau. A *minimal surface* is a surface Σ in the Euclidean 3-space $\mathbf{E}^3$ which locally minimizes area. That description is mathematical shorthand for saying that if you begin with the surface Σ, focus your attention on a small portion $\Omega \subset \Sigma$ of it, and then perturb the position of Σ only on Ω and just a little bit, then the resulting area of the perturbed surface will not be lower — more precisely, thinking of the perturbation as resulting in a family Σ_t of surfaces beginning with the original surface Σ (now called Σ_0), we require that the time t derivative at $t = 0$ of the area of Σ_t should vanish. See Figure 1.

Flat structures for minimal surfaces

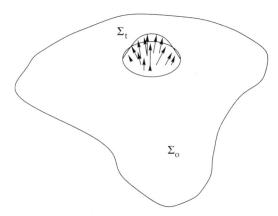

Figure 1. Deformation Σ_t of a surface Σ_0.

Here are some examples. In most mathematical subjects, there is a trivial example, and the subject of minimal surfaces is no exception: any portion Σ of a plane in $\mathbf{E}^3$ is minimal, because if you perturb a small neighborhood on a plane, then the projection of that perturbation back onto the plane pretty clearly both lowers the area of the perturbation and provides a comparison with the area of the deformed planar region.

For the next examples, let's first assert that portions of soap films are good models of minimal surfaces, up to negligible effects of the downward drag of gravity, the thickness of the frame off of which the film hangs, some physical properties of the soap solution, etc. — see the wonderful book [**HT85**] by Hildebrandt and Tromba for a careful and gentle discussion.

But if you accept that intuition, then you can begin to visualize minimal surfaces through some imagination experiments. For instance, imagine taking two circular pieces of wire, holding them parallel to each other (one just above the other) and dipping the pair into a bucket of soapy water (if you ever do this for school children, use 10 parts water to 1 part Joy or Dawn and then add a few tablespoons of glycerine for very sturdy bubbles). When you pull the pair of circles out of the bucket, you'll find the pair spanned by a curved cylinder

— from the point of view of imagination, the cylinder wouldn't have flat walls, because one could save some area by slightly pulling the cylinders towards the center of the circle. Later on, I will convince you that this film is idealized by a surface of revolution (that part is maybe evident already) called a catenoid, generated by the profile curve $Z = \cosh X$ in (X, Y, Z) space. See Figure 2.

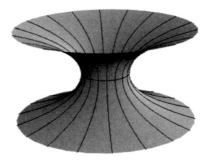

Figure 2. The catenoid.
Figure by Matthias Weber.

In fact, our primary interest today is in *complete* minimal surfaces, i.e., those on which there are no arcs of finite length on the surface that leave the surface; informally, such surfaces Σ extend indefinitely.

As usual, the first example is pretty trivial — it is just an entire plane in $\mathbf{E}^3$; note that any curve on Σ that starts at a point of Σ and leaves any compact set it enters must have infinite length. The next example is found via a thought experiment as follows. Take a pair of circles and dip them in a soap solution as before to make a catenoid. Now take a pair of circles of larger radius, hold them further apart and dip them in a soap solution to make a bigger catenoid.

Flat structures for minimal surfaces

You should be able to find a distance at which to hold them so that the old smaller catenoid fits exactly as a small piece of the new larger catenoid. Continue this process of fitting larger catenoids to extend the smaller catenoids: the distances between the bounding (huge) circles grow increasingly (and arbitrarily) large, and the limiting object has no boundary at all. Mathematically, it is the entire surface of revolution of the profile curve $Z = \cosh X$ in (X, Y, Z) space.

In Figure 3, all arcs on the surface have infinite length. Here you regard the surface as continuing in its pattern in all horizontal directions. This is actually the famous doubly periodic minimal surface of Scherk. Here, you should imagine leaving the picture in three ways: either vertically by going up or down along the nearly vertical almost planar regions, or horizontally, by following the diagonal straight lines out, or by weaving your way around the curvy pieces. Incidentally, if you were to look straight down onto this surface from above, you'd see an infinite checkerboard.

Figure 3. Doubly periodic minimal surface of Scherk.
Figure by Matthias Weber.

In Figure 4, you see a different sort of example: here the arc Γ_1 leaves the region in finite length, and the arc Γ_2 falls into a hole in finite length.

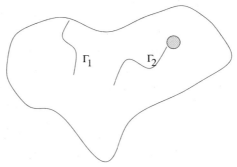

Figure 4. On this surface, the arcs Γ_1 and Γ_2 are of finite length.

2. Some history

The subject of minimal surfaces was one of the earliest areas of study of modern geometry, with the catenoid being discovered in the 18th century. In that century, a third (in addition to the plane) minimal surface was discovered — it is called the helicoid and is described as follows. Take a ruler and hold it at arm's length, parallel to the ground. Rotate the rod (still parallel to the ground) and begin raising your arm (and rod). If you manage to both raise your arm and rotate the rod at constant rates, then the surface which the rod is describing is a helicoid. From this description, of course, you can easily find a precise parametrization of the helicoid. See also Figure 5.

I am describing these surfaces so carefully because of an astonishing piece of mathematical history. For two centuries, minimal surfaces were intensively studied by differential geometers, and those two examples of the plane and the catenoid were the only examples known which were complete without self-intersections (it turns out to be pretty easy to produce examples with self-intersections) and which

Flat structures for minimal surfaces

didn't repeat (the helicoid repeats.) Some mathematicians took this historical failure of attempts to find different examples as evidence

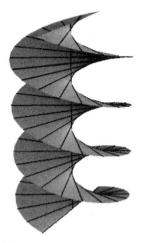

Figure 5. The helicoid.
Figure by Matthias Weber.

Figure 6. Costa's surface.
Figure by Matthias Weber.

that there were no additional examples, and indeed there are a number of proven obstructions to the existence of examples. Yet here's the amazing part: in 1982, a Brazilian graduate student named Celso Costa [**Cos84**] produced a new example of a complete, embedded minimal surface in $\mathbf{E}^3$. It also had other properties; for example, its group of self-symmetries was finite. But most remarkable, and what makes it natural for its discovery to have taken so, long is that it was topologically non-trivial. While the helicoid and the plane are topological disks, and the catenoid is topologically but a punctured disk, Costa's surface was a thrice-punctured *torus*. See Figure 6. Costa's surface changed the perspective we have on this subject: now we know that there are many more complicated but geometrically nice complete minimal surfaces in $\mathbf{E}^3$ than we suspected twenty years ago, and we can hope to find ways to classify them. This is a subject very much in its infancy.

3. The Weierstrass representation

In this chapter, we are aiming to prove the existence of some minimal surfaces in space. There are roughly two standard methods, the first from a real analysis perspective, and the second from a complex analytic, almost algebraic, point of view. Let me describe the first method very vaguely, almost as a cartoon. My goals in doing this are first to expose you, ever so slightly, to a deep and powerful area of mathematics, and second to provide some contrast to the second method on which I will focus for the rest of this talk; most of all, I include this discussion because omitting it would be horribly misleading to a student who might be meeting the subject of the calculus of variations for the first time.

3.1. Direct method. With those caveats done, here is a method from real analysis. You want to prove the existence of a minimal surface in space that "looks like" a certain surface $\Sigma \subset \mathbf{E}^3$ that you have in mind. Okay, so consider "the space $\mathcal{B} = \{\Sigma\}$ of surfaces Σ that are like Σ_0." Obviously, it is a non-trivial matter to clarify that intuition so as to define the space $\mathcal{B}$ properly. But let's pretend that we've done that (and like this whole approach, this defining

Flat structures for minimal surfaces

process has been done successfully many times in the course of proving remarkable results), and we've even endowed $\mathcal{B}$ with a topology (so that we know when two surfaces $\Sigma \in \mathcal{B}$ and $\Sigma' \in \mathcal{B}$ are "close"). Since each surface $\Sigma \in \mathcal{B}$ has an area, or possibly a modified area, we can then consider a sequence $\langle \Sigma_n \rangle \subset \mathcal{B}$ so that the areas (Σ_n) are tending to a minimum or a critical point. Then, we attempt to prove that this sequence has a subsequence which converges nicely, in some sense, to a nice limit. Good places where you might read about this method are [**Mo85**] and [**Str88**].

3.2. Intrinsic vs. extrinsic geometry. The method we will focus on comes from complex analysis and goes by the name "Weierstrass representation". I will summarize this method in a moment, but first let me distinguish for you between two types of geometry; begin by noting that a surface $\Sigma \subset \mathbf{E}^3$ has a geometry of lengths and angles that comes from how it sits in $\mathbf{E}^3$. If we then remove Σ from $\mathbf{E}^3$ and study this geometry of Σ without further reference to how it sits in $\mathbf{E}^3$, we are studying its *intrinsic geometry*. If we also want to study the embedding of $\Sigma \subset \mathbf{E}^3$, for instance, how it curves in space, then we are studying its *extrinsic geometry*. For example, because you can roll a paper towel onto a tube, as well as lay it flat, a portion of a plane and a portion of a cylinder have the same intrinsic geometry but different extrinsic geometry. It's a bit harder to see that portions of a catenoid have the same intrinsic geometry as portions of helicoids (but they do), but it is easy to see that they have different extrinsic geometry. To formally prove that the extrinsic geometries are different, you might observe that the catenoid contains no straight lines, while the helicoid is a union of straight lines. On the other hand, the map which identifies the intrinsic geometry of a portion of the helicoid with the intrinsic geometry of the catenoid takes one of the defining straight lines of the helicoid to a planar curve in the catenoid which connects its two ends.)

With that distinction made, we can say (glibly) that the Weierstrass representation method for producing minimal surfaces is just to write them down explicitly; a more careful summary is that we understand the intrinsic geometry well enough to be able to create functions corresponding to the extrinsic geometry so that an explicit minimal (local) embedding may be found. This is a very useful perspective to take when seeking critical points $\Sigma \subset \mathcal{B}$ for area which are *not* local minimizing; you can see from the chapter of Robin Forman how easily a minimizing sequence Σ_n for a Morse function can "pass right by" critical points on a manifold M which are not local minima, and the same phenomenon holds at least equally well for spaces such as $\mathcal{B}$. That is a very vague outline of the point-of-view the Weierstrass representation school takes; we'll now clarify this by precisely describing the Weierstrass representation. We have assumed very little background in real differential geometry or complex analysis, so this will take some time.

3.3. Riemann surfaces. One can view a surface $\Sigma \subset \mathbf{E}^3$ in space as having either a very flabby structure or a very tight structure, depending on the context in which you encounter the surface. That is pretty vague, but what I mean is that if you are studying topology, then all homeomorphic surfaces in space may be equivalent for you, and certainly small perturbations of the surface are immaterial for you. If you are a Riemannian geometer, then those slight perturbations change the surface entirely for you, as you only identify surfaces $\Sigma, \Sigma' \subset \mathbf{E}^3$ if there is maybe a congruence between them, or at least a way of matching up their intrinsic distance and angle functions.

There are many notions of middle ground between these two extremes, and one is particularly important for our study of minimal surfaces. We will regard two surfaces, $\Sigma, \Sigma' \subset \mathbf{E}^3$ as equivalent if there is a way of identifying them, say by a diffeomorphism $\varphi : E \to E'$ which preserves angle measurements. This is, of course, stronger than just requiring the surfaces to be homeomorphic, but it also is weaker than requiring that both the angles and the lengths agree. For instance, if we begin with a catenoid $C \subset \mathbf{E}^3$ and then replace each point $(X, Y, Z) \in C$ with a new point $(7X, 7Y, 7Z)$ to get a new catenoid $C' = \{(7X, 7Y, 7Z) \mid (X, Y, Z) \in C\}$, then the map $S : C \to C'$ which takes $(X, Y, Z) \mapsto (7X, 7Y, 7Z)$ distorts distances by a factor of 7, but preserves angles. This is the most trivial kind of angle-preserving, length-destroying transformation, but there are many others.

Flat structures for minimal surfaces 89

A surface endowed with a notion of angle at each point (or if you prefer, with an equivalence class of (Riemannian) distance functions, where the equivalence requires preservation of angles) is called a "Riemann surface".

It turns out that to define a minimal surface, one only needs to consider the Riemann surface underlying the minimal surface; the intrinsic geometry relating to measures of distances is largely irrelevant.

3.4. Some real differential geometry. We are interested in characterizing minimal surfaces, i.e., surfaces for which small perturbations of small subsets of the surface result in a new perturbed surface whose area is at least as large as the area of the original surface. Because we have so many choices of sets on which to perturb and allowable perturbations on that set, this condition is a geometric condition at each point of the surface. What should that condition be? Well, part of the power of the well-developed machinery of the calculus is that you never need to think about that answer — you can just compute the Euler-Lagrange equation for the area functional, and you can find the answer in the article of Frank Morgan in this volume.

But of course, if you choose, you can think about it and reason out the answer by geometrical arguments. For instance, try the thought experiment: can a minimal surface have a portion that is concave down like the arctic cap of the world? Certainly not, because we could always reduce the area of that cap by flattening a portion of the cap: a planar disk with the same boundary as the cap has less area than the cap. A slightly more subtle example in the same direction is that of a high mountain ridge, such as you might get by balancing a pillow along an edge. (Mathematically, consider the graph $\{(x,y,z) \mid z = 5 - x^2\}$.) Here we can reduce the area of the ridge by pushing down slightly in the center; here we increase the length of the ridge line (the line $\{(0, y, 5)\}$ in the example) slightly but decrease the lengths of all of the orthogonal directions to more than compensate, so that the *total* area declines.

By a few more experiments along these lines, you can conclude (and this does take some thought, so don't be discouraged if it takes you some time to see it) that the condition for minimality is that, informally, the surface bends up as much as it bends down. Thus every point of the surface looks like the seat of a perfectly balanced horse saddle, with the curvature of the curve from the horse's head to its tail being equal (but opposite in direction) to the curvature of the orthogonal curve from the left side of the horse over its back to the right side.

Let's be a bit more formal. In this paragraph, I'll assume you remember from your multivariable calculus class the meaning of the curvature $\kappa(\gamma)(p)$ of a plane curve γ at a point $p \in \gamma$: it is the reciprocal of the radius $r(C_p)$ of the osculating circle C_p for γ at p,

$$\kappa(\gamma)(p) = r(C_p)^{-1}.$$

Now for a surface $\Sigma \subset \mathbf{E}^3$, pick a point $p \in \Sigma$ and consider a normal $\vec{n}$ to Σ at the point $p \in \Sigma$. There is a whole circle's worth of planes through Σ at p that contain $\vec{n}$, and each one, say P_θ (where θ denotes a point in the parametrizing circle), meets Σ in a planar arc, say γ_θ, through p. Let κ_θ denote the (signed) curvature of γ_θ at p, where the sign of κ_θ is positive if the normal $\vec{n}$ points into the osculating disk and negative if the normal $\vec{n}$ points out of the osculating disk. There is some angle θ, say $\theta_{\max}$, at which the curvature function κ_θ is least, and another angle θ, say $\theta_{\min}$, at which the curvature function κ_θ is most.

You might guess, at this point, that the two curves $\gamma_{\theta_{\max}}$ and $\gamma_{\theta_{\min}}$, along which the curvatures have the maximum $\kappa_{\max}$ and the minimum $\kappa_{\min}$, are orthogonal; this is true, and I invite you to prove it, both geometrically and analytically.

The *mean curvature* $H(p)$ of Σ at p is described by

$$H(p) = \frac{1}{2}\left(\kappa_{\max} + \kappa_{\min}\right).$$

(Recall that given a curve γ in the plane, there is a unique circle that is both tangent to the curve and has the same second derivative: by this I mean that if we travel along γ away from p a distance ϵ, then the circle is at a distance proportional to ϵ^3. This circle is called the osculating circle C_p for γ and p.) Then the curvature $\kappa(\gamma)(p)$ and our previous thought experiments on area have led us to the conjecture

Flat structures for minimal surfaces

that a minimal surface must satisfy the condition that $\kappa_{\max} = -\kappa_{\min}$, i.e., that
$$H(p) = 0$$
for each point p on a minimal surface.

A word of warning: while it is true that $\kappa_{\max} = -\kappa_{\min}$ are in balance, we are not saying that $\kappa_{\max}$ (or $\kappa_{\min}$) is constant. The beauty of soap film lies in the curvatures being balanced at every point, *and* changing from point to point.

In fact, a stronger statement is true. Suppose you have a surface Σ and a function φ on Σ which vanishes outside a small set of Σ. Then you can consider a (continuous) perturbation $\Sigma_{t,\varphi}$ of Σ via the formula (using vector addition in $\mathbf{E}^3$)
$$\Sigma_{t,\varphi} = \Sigma + t\varphi(\vec{n})$$
where $\vec{n}$ is a choice of unit normal field on Σ, and t is a small time parameter varying in an interval through time $t = 0$ (corresponding to $\Sigma = \Sigma_{0,\varphi}$). You can ask how the area $\text{Area}(\Sigma_{t,\varphi})$ changes with t, and we get an answer for the first derivative of
$$\left.\frac{d}{dt}\right|_{t=0} \text{Area}(\Sigma_{t,\varphi}) = -\iint_\Sigma \varphi H \, d\,\text{Area}$$
where H is the mean curvature function defined in the last paragraph. (This is a pretty straightforward computation, of the sort described in the Frank Jones chapter: we move the $\frac{d}{dt}$ operator into the integral sign to get an integral involving derivatives along the surface of the deformation function φ. The standard method is then to integrate by parts to end up with an integral (against some function built out of derivatives of the original surface ϵ_0). From this formula, you see that a surface is minimal if and only if its area doesn't change to first order under any perturbation and that holds if and only if the mean curvature H of Σ vanishes entirely on Σ.)

3.5. Digression on complex analysis. It is, of course, obvious that these surfaces we are talking about here are two-dimensional, i.e., an ant crawling around on them thinks he is on some warped version of a piece of $\mathbf{R}^2$. But, of course, it is often very convenient to regard the

2-plane $\mathbf{R}^2$ as the complex numbers $\mathbf{C}$: instead of writing coordinates for a point $p \in \mathbf{R}^2$ in terms of a pair (x,y) of real numbers x and y, we write the coordinates for $p \in \mathbf{R}^2$ as a *single complex* number $z = x + \sqrt{-1}y = x + iy$.

I do want to use a bit from the subject of complex analysis here. If you've had a course in that, then you're well-prepared for my remarks. If not, then resolve to take one — the mathematics you'll learn is both beautiful and, coincidentally, fundamental — and here are a few remarks to help you follow and appreciate the rest of the chapter.

Of course you know that a function f on $\mathbf{R}^2$ (either real-valued with $f : \mathbf{R}^2 \to \mathbf{R}$, or complex-valued with $f : \mathbf{R}^2 \to \mathbf{C}$) can be written as $f = f(x,y)$. A little bit of algebra will convince you that we might also write $f = f(z,\bar{z})$ where $z = x + iy$ and the complex conjugate $\bar{z}$ of z is written $\bar{z} = x - iy$. This makes sense as $x = \frac{1}{2}(z+\bar{z})$ and $y = -\frac{i}{2}(z-\bar{z})$, using that $-i(i) = -i^2 = -(-1) = 1$. Now complex analysis begins by focussing on the study of functions $f = f(z,\bar{z})$ which, informally, do not depend on $\bar{z}$.

This requires some explanation. The basic point is that we want to be able to define the derivative of f in the way in which we've been accustomed since calculus, i.e.,

$$(1) \qquad f'(z) = \lim_{z \to a} \frac{f(z) - f(a)}{z - a}$$

where here we compute subtraction and division according to the natural rules for complex numbers that came out of the definition $i^2 = -1$. Try a couple of examples, such as $f(z) = \bar{z}$ or $f(z) = (\operatorname{Re} z)^2$, and you'll see that (1) is a very strong condition: for instance, if we take $f(z) = \bar{z} = re^{-i\theta}$ in the natural notation, then the difference quotient at zero will be

$$\frac{f(z) - f(0)}{z - 0} = \frac{\bar{z}}{z} = \frac{re^{-i\theta}}{re^{i\theta}} = e^{-2i\theta}$$

so that as $z \to 0$, the limit depends on the direction of approach to zero.

This is pretty unsatisfying, since we'd like a complex derivative of a function f on $\mathbf{C}$ to be a function on $\mathbf{C}$ itself, assigning to each point of $\mathbf{C}$ a single, well-defined complex number and not some family of

Flat structures for minimal surfaces

values depending on paths of approach. So we need a rule that assigns to each point $z \in \mathbf{C}$ a complex number $f'(z)$. Let's require the limit (1) to exist everywhere; this rules out the vast majority of smooth functions $f : \mathbf{R}^2 \to \mathbf{R}^2$. Indeed, the only functions that remain satisfy a pair of differential equations on their real ($\operatorname{Re} f = \frac{1}{2}(f + \bar{f})$) and imaginary ($\operatorname{Im} f = -\frac{i}{2}(f - \bar{f})$) parts:

$$\frac{\partial}{\partial x}(\operatorname{Re} f) = \frac{\partial}{\partial y}(\operatorname{Im} f)$$
(2)
$$\frac{\partial}{\partial y}(\operatorname{Re} f) = -\frac{\partial}{\partial x}(\operatorname{Im} f).$$

These equations, like most new equations, are a bit daunting at first, but you get used to them. Here's a way to simplify the presentation somewhat. Consider the differential operators

$$\frac{\partial}{\partial z} = \frac{1}{2}\left(\frac{\partial}{\partial x} - i\frac{\partial}{\partial y}\right)$$
(3)   and
$$\frac{\partial}{\partial \bar{z}} = \frac{1}{2}\left(\frac{\partial}{\partial x} + i\frac{\partial}{\partial y}\right).$$

(Yes, I got the signs right: note that $\frac{\partial}{\partial z}z = \frac{\partial}{\partial \bar{z}}\bar{z} = 1$ while $\frac{\partial}{\partial \bar{z}}z = \frac{\partial}{\partial z}\bar{z} = 0$.) Then the equations (2) are just the real and imaginary parts of the equation

(4)
$$\frac{\partial}{\partial \bar{z}}f = 0,$$

and this justifies the remark that complex analysis is, at the outset, a study of functions $f = f(z, \bar{z}) : \mathbf{C} \to \mathbf{C}$ that depend only on z and not on $\bar{z}$.

In fact, (2) or equivalently (4) is a very restrictive system of partial differential equations, and the functions that satisfy (2) in a domain Ω, called holomorphic functions on Ω or (complex) analytic functions on Ω have many interesting and surprising properties. In particular, all such functions admit a representation in terms of a convergent power series in the variable z, the real and imaginary parts of a holomorphic function are harmonic functions (i.e., satisfy $\Delta \operatorname{Re} f = \left(\frac{\partial^2}{\partial x^2} + \frac{\partial^2}{\partial y^2}\right)\operatorname{Re} f = 0$), and the path integrals $\int_\gamma f(z)dz$ depend only on the endpoints of γ, and not on the particular path

γ between the endpoints. We will have more to say about the last two of these properties a bit later. Let me just append one additional piece of terminology: a function f is called *meromorphic* if it is a holomorphic map to $\widehat{\mathbf{C}} = \mathbf{C} \cup \{\infty\}$. This means that the function f takes on the value ∞ at some isolated places $\{z_i\}$, and at only those places, there is a well-defined infinite limit $\lim_{z \to z_i} f(z) = \infty$ (or, equivalently for the topology of $\widehat{\mathbf{C}} = \mathbf{C} \cup \{\infty\}$, we have that $\lim_{z \to z_i} |f(z)| = \infty$). These points z_i are called the *poles* of f.

3.6. From real differential geometry to complex analysis. I want to take a *parametric* approach to this minimal surface problem. You see, all along we have been talking about a surface $\Sigma \subset \mathbf{E}^3$ contained in $\mathbf{E}^3$ and how $\mathbf{E}^3$ induces on this surface a way of measuring angles. Now I want to separate psychologically the surface Σ with its *intrinsic* notion of angles (called a conformal structure) from the way it lies within $\mathbf{E}^3$. So now imagine a surface Σ that comes equipped with a well-defined notion of angle: this makes Σ into a *Riemann surface* — we will change the way we denote the surface from Σ to $\mathcal{R}$ to emphasize that the surface has this extra structure of angles on it. Also imagine a mapping $u : \mathcal{R} \to \mathbf{E}^3$ so that the image set $u(\mathcal{R})$ lands on Σ and the way we measure angles on $\mathcal{R}$ at a point $p \in \mathcal{R}$ agrees with the way Euclidean space $\mathbf{E}^3$ forced us to measure angles at $u(p)$ on $\Sigma = u(\mathcal{R})$. The mathematical terminology that describes this situation is that the map u is a *conformal* (i.e., angle-preserving) map of the Riemann surface $\mathcal{R}$ onto Σ.

Now, since $\mathcal{R}$ has a way of measuring angles, we observe that there is a natural way of rotating a vector by $90°$ to another vector. It turns out that in 2 real dimensions, the fact that we have a continuous way of rotating vectors by $90°$ implies that there is a consistent way to regard open sets on $\mathcal{R}$ as open sets in $\mathbf{C}$. In other words, we can define a notion of (complex) analytic function on the surface; i.e., a Riemann surface is a valid domain for complex analysis.

[This is actually a strong statement: imagine two tori of revolution, one, say T_a, generated by revolving the circle $(x-1)^2 + y^2 = a^2$ and one, say T_b, generated by revolving the circle $(x-1)^2 + y^2 = b^2$ around the z-axis in $\mathbf{E}^3$. (Now we cannot talk about holomorphic

Flat structures for minimal surfaces 95

functions on these tori, as it turns out that any globally holomorphic function on a compact boundaryless Riemann surface is a constant; but we can talk about *meromorphic* functions on T_a and T_b, i.e., functions which are holomorphic off of a set of punctures and whose absolute value goes to infinity as we tend towards the puncture.) It turns out that there is no angle-preserving way of mapping the first torus homeomorphically onto the second torus, so the space of meromorphic functions on T_a is quite distinct from the space of meromorphic functions on T_b.]

Let's return to our map $u : \mathcal{R} \to \mathbf{E}^3$. Now, there is usually no way to put a global coordinate system in a surface: if you think about a torus, for instance, it is likely that if you were to follow the x-axis of a coordinate system far enough out, you would end up returning to near where you started. So how can we do calculus or complex analysis on a surface? We would seem to need to be in a piece of the plane for that. Actually, that's the answer: we imagine the surface being made up of "patches" of the plane, and we are careful to be sure that no answer we obtain depends critically on what patch we use. We will now proceed with one such piece of analysis on a patch. Begin with the understanding that patches on $\mathcal{R}$ can be identified with patches on $\mathbf{C}$, so it is useful psychologically to confuse the two patches and think of having a complex variable z on a patch of $\mathcal{R}$. As usual, write $z = x + iy$, and as long as we're setting notation, let $\vec{u} = (X, Y, Z)$. (In other words, on a small bit of $\mathcal{R}$, we impose (x, y) coordinates written as $z = x + iy$, and we regard u as assigning to z a position in space: here space has the standard coordinates (X, Y, Z), and so u can be written concisely as $u : z \to u(z) : (X(z), Y(z), Z(z))$.) We translate the geometric condition that the map u is angle-preserving to the equivalent analytic conditions that

(5a)
$$\left| \frac{\partial \vec{u}}{\partial x} \right| = \left| \frac{\partial \vec{u}}{\partial y} \right|$$
and
$$\frac{\partial u}{\partial x} \cdot \frac{\partial u}{\partial y} = 0.$$

As geometers, we like to compute geometric quantities; indeed, since our choice of patch is arbitrary, we cannot really trust answers or operators in coordinates that we can't explain in a geometric way without coordinates. It is always interesting in geometric analysis to find the Laplacian of natural geometric quantities, and now is not an exception. So we ask, what is $\Delta \vec{u}$ equal to?

Well, $\Delta = \frac{\partial^2}{\partial x^2} + \frac{\partial^2}{\partial y^2}$ is a natural operator because, as it is invariant under rotation of coordinates $(x, y) \mapsto (x^*, y^*)$, it is geometric in a strong sense. Thus, $\Delta \vec{u}$ is geometric in a strong sense: our result should be a vector, and one whose coefficient is both geometric and a sum of second derivatives in orthogonal directions. The point is that one can pretty much guess the answer with some thought, or with less thought recognize that the actual answer that

$$\Delta \vec{u} = -2H\overrightarrow{N},$$

where $\overrightarrow{N}$ is the normal vector, is not a great surprise; there are no other vectors naturally connected with $u(\mathcal{R})$ other than $\overrightarrow{N}$ and no natural functions associated to an embedding of a patch by $\vec{u}$ which involve the required sum of second derivatives other than the mean curvature function H.

Of course, for a minimal surface, we know that $H \equiv 0$, as this is the defining property for minimal surfaces. We conclude that we have two conditions for our conformal minimal immersion $\vec{u} : \mathcal{R} \to \mathbf{E}^3$ of the Riemann surface $\mathcal{R}$: u must be angle-preserving as in (5a) and

(5b) $$\Delta \vec{u} = 0.$$

Note that if we write the map $u : \mathcal{R} \to \mathbf{E}^3$ in its $\mathbf{E}^3$ coordinates, i.e., $u = (X, Y, Z)$, then (5b) is really three equations. It's convenient to assign the three equations the same label:

(5b)
$$\Delta X = 0,$$
$$\Delta Y = 0,$$
$$\Delta Z = 0.$$

Now let's bring complex analysis into our discussion. Using the complex differential operators $\frac{\partial}{\partial z}$ and $\frac{\partial}{\partial \bar{z}}$ defined in (3), we can rewrite the Laplacian as

$$\Delta = \frac{\partial^2}{\partial x^2} + \frac{\partial^2}{\partial y^2} = 4\frac{\partial}{\partial \bar{z}}\frac{\partial}{\partial z},$$

Flat structures for minimal surfaces 97

and then (5b) can be written as

(6b)
$$\frac{\partial}{\partial \bar{z}}\left[\frac{\partial}{\partial z}X\right] = 0,$$
$$\frac{\partial}{\partial \bar{z}}\left[\frac{\partial}{\partial z}Y\right] = 0,$$
$$\frac{\partial}{\partial \bar{z}}\left[\frac{\partial}{\partial z}Z\right] = 0.$$

The wonderful fact is that (5a) can *also* be rewritten in a very clear way in terms of the same bracketed objects in (6b), as

(6a)
$$\left(\frac{\partial}{\partial z}X\right)^2 + \left(\frac{\partial}{\partial z}Y\right)^2 + \left(\frac{\partial}{\partial z}Z\right)^2 = 0.$$

Look at (6a) and (6b) together: the set of equations (6b) says that the objects $\frac{\partial}{\partial z}X$, $\frac{\partial}{\partial z}Y$, and $\frac{\partial}{\partial z}Z$ are all holomorphic as (6b) is the defining equation (compare (4)) for holomorphicity, and (6a) is an *algebraic* relation among those three terms.

Through all of this work on finding and rewriting necessary conditions for the minimal mapping $u : \mathcal{R} \to \mathbf{E}^3$, don't lose sight of the goal of writing down a map u, preferably in coordinates as $u = (X, Y, Z)$. Keeping this goal in mind, notice how close we are to finding a solution: the quantities for which we are now writing conditions, i.e., $\frac{\partial X}{\partial z}$, $\frac{\partial Y}{\partial z}$, $\frac{\partial Z}{\partial z}$ are just the derivatives of those coordinate functions we seek. In particular, if we find nice solutions $A = \frac{\partial X}{\partial z}$, $B = \frac{\partial Y}{\partial z}$ and $C = \frac{\partial Z}{\partial z}$, then we can recover the map $u = (X, Y, Z)$ by just integrating as in first semester calculus:

(7)
$$X(w) = \operatorname{Re} \int_p^w A\,dz = \operatorname{Re} \int_p^w \frac{\partial X}{\partial z},$$
$$Y(w) = \operatorname{Re} \int_p^w B\,dz = \operatorname{Re} \int_p^w \frac{\partial Y}{\partial z},$$
$$Z(w) = \operatorname{Re} \int_p^w C\,dz = \operatorname{Re} \int_p^w \frac{\partial Z}{\partial z}.$$

In fact, if all we want to do is minimally embed a small disk $D \subset \mathcal{R}$ on $\mathcal{R}$, we almost just need to find three holomorphic functions A, B, and C on D with the property (6b) that $A^2 + B^2 + C^2 = 0$ on

D, and then use (7) to write the map $u : D \subset \mathcal{R} \to \mathbf{E}^3$. Let me defer discussion of the word "almost" until section 3.9.

I was careful in that last paragraph to work within a small disk $D \subset \mathcal{R}$. Why is this? The reason is that I have not been paying close attention so far to what "tensor type" the A, B, and C should be.

This requires some explanation — especially if you find the word "tensor" intimidating. Basically, in geometry we are interested in *geometric* objects where by this I have in mind not only functions but vector fields, line elements (i.e., objects you integrate along curves to assign a length to the curve), area elements (i.e., objects you integrate over sets to assign an area to a set), etc. These are natural objects to picture (especially, e.g., vector fields), but how do we compute with them (especially on shapes such as the sphere or the torus on which it's impossible to impose a single global coordinate system)? When you write down such an object in coordinates, its expression depends on coordinates: for instance, the vector field on the plane consisting of unit eastwardly pointing vectors might be written as $(1, 0)$ everywhere, but if you change coordinates from (x, y) to $(u, y) = (2x, y)$, then in terms of (u, y), you would want to write that field as $(u, y) = (2, 0)$. The idea, then, of tensors is just to write out a geometric object in coordinates together with an explanation of how the expression changes when you change coordinates. (This is sometimes a bit much to swallow on the first reading. For the sake of correctness of exposition, I am going to proceed writing as if you've understood all of the above. If, however, this is your first encounter with tensors, and you're still not comfortable with them, then please don't get stuck here: I don't think you'll lose too much just replacing all of the tensoral language such as "one-forms" with mental language such as "geometric objects that are locally like functions that it makes sense to integrate." I've been focussing just on the algebra and contenting myself to work within a small disk $D \subset \mathcal{R}$ where I can just take A, B, C to be functions — this prevents me from getting into any trouble if they happen to be of a different tensor type. Looking at them more carefully though, you observe that, for instance, we wrote $A = \frac{\partial X}{\partial z}$, and since z is a *choice* of coordinate on a domain $\Omega \subset \mathcal{R}$, we need to consider how A would transform if we were to choose a

Flat structures for minimal surfaces

different coordinate, say ζ, on the domain Ω. Imagine, for example, how A would transform if we were to choose $\zeta = 5z$. In particular, we want the coordinate $X = \operatorname{Re} \int A\,dz$ to remain invariant under such a change of coordinate from z to ζ. The solution to these anxieties is simply to note that we should really be imagining A as the coefficient of a "one-form α"; i.e., in coordinates,

$$\alpha = A\,dz = \frac{\partial X}{\partial z}dz.$$

The "dz" on the end tells you how to change coordinates from, say, z to ζ: since the chain rule says that $dz = \frac{dz}{d\zeta}d\zeta$, we see that when we rewrite $\alpha = A\,dz$ in terms of a different coordinate ζ, we should write it as $\alpha = A\,dz = A(\frac{dz}{d\zeta}) = (A\frac{dz}{d\zeta})d\zeta$; i.e., the A transforms to $A\frac{dz}{d\zeta}$ under the change of coordinates $z \mapsto \zeta$. Why is this important? Well, now not only will α transform properly under a change of coordinates from z to ζ, but then also the expression $\operatorname{Re} \int \alpha$ will make sense with respect to any choice of coordinate on $\mathcal{R}$. (Generally speaking, "one-forms" are dual to vector fields: when you apply a one-form at a point to a vector at that point, you just get a number. Sometimes I even like to say that one-forms are the geometric objects that eat vectors and return numbers.)

Let's summarize: we require (i) "meromorphic" one-forms α, β, γ, say (in coordinates)

(7b)
$$\alpha = \frac{\partial X}{\partial z}dz,$$
$$\beta = \frac{\partial Y}{\partial z}dz,$$
$$\gamma = \frac{\partial Z}{\partial z}dz,$$

with (ii) the property that their squares (often called "quadratic differentials") α^2, β^2 and γ^2 sum to zero:

(7a)
$$\alpha^2 + \beta^2 + \gamma^2 = 0.$$

3.7. The Weierstrass representation. Let's solve the pair of equations (7a) and (7b). Of paramount importance in our solution to these equations is that our solution be both (i) simple and (ii) geometric. Here's the solution from the 1860s called the Weierstrass-Enneper representation: find a meromorphic function G and a meromorphic one-form dh (not necessarily exact, but more on that later) and set

(8)
$$\alpha = \frac{1}{2}\left(\frac{1}{G} - G\right)dh,$$
$$\beta = \frac{i}{2}\left(\frac{1}{G} + G\right)dh,$$
$$\gamma = dh.$$

Of course, since G and dh are meromorphic, these forms α, β and γ are also meromorphic, and then it is also quite easy to check that $\alpha^2 + \beta^2 + \gamma^2 = 0$. The expression $\alpha^2 + \beta^2 + \gamma^2$ *is* quadratic, so the expression (8), while maybe appearing a bit complicated at first, is about as simple as one could hope for. But where is the geometry? First off, remember how everything fits together: the map $u : \mathcal{R} \to \mathbf{E}^3$ is now given as

(9)
$$z \mapsto \left(\operatorname{Re}\int_p^z \alpha, \operatorname{Re}\int_p^z \beta, \operatorname{Re}\int_p^z \gamma\right)$$
$$= \left(\operatorname{Re}\int_p^z \frac{1}{2}\left(\frac{1}{G} - G\right)dh, \operatorname{Re}\int_p^z \frac{i}{2}\left(\frac{1}{G} + G\right)dh, \operatorname{Re}\int_p^z dh\right)$$

(where $p \in \mathcal{R}$ is just some arbitrarily chosen point that will map to $(0,0,0) \in \mathbf{E}^3$). Look now at the last coordinate, $\operatorname{Re}\int dh$. We see that $Z = \operatorname{Re}\int dh$, so that by taking the derivative of Z, we see that the real form dZ can be expressed as $dZ = \operatorname{Re} dh$ (by the fundamental theorem of calculus), and so dh is obtained by taking dZ and then "complexifying" the one-form (adding $\sqrt{-1}$ times the *conjugate* differential, for those who have studied a semester of complex analysis). Informally, we think of dh as d"height", and this explains its geometric meaning.

The function G is maybe a bit more subtle. Remember that on the surface $\Sigma = u(R)$, there is a single naturally associated vector, the normal vector $\vec{N}$. We can then think of a function which associates, to a given point p on Σ, a unit vector $\vec{N}(p)$. (Really, there are two choices, the first being the negative of the second, but we can just pick one at a point P_0, and if the surface Σ is orientable and if we assume that $\vec{N}(p)$ is continuous on Σ, then we'll have completely

Flat structures for minimal surfaces

determined the choice of $\vec{N}$ everywhere.) What is the range of this function $\vec{N}$? Well, it associates a unit vector to a point $p \in \Sigma$, so the range is the unit sphere $S^2 = \{(X,Y,Z) \in \mathbf{E}^3 \mid X^2 + Y^2 + Z^2 = 1\}$. There is a natural map, called stereographic projection $S : S^2 \to \mathbf{C}^2 = \mathbf{C} \cup \{0\}$, that connects the surface S^2 to complex analysis. The construction (see Figure 7) is as follows: put the unit sphere S^2 on top of the complex plane $\mathbf{C}$, with the south pole S sitting at the origin. Then shoot off a ray in $\mathbf{E}^3$ from the north pole n to another point $q \in S^2$ on S^2. The ray continues past q down to the plane $\mathbf{C}$, meeting $\mathbf{C}$ at a unique point $S(q) \in \widehat{\mathbf{C}} = \mathbf{C} \cup \{0\}$. The process is completely reversible, and so the map $S : S^2 - \{n\} \to \mathbf{C}$ is an identification of $S^2 - \{n\}$ with $\mathbf{C}$. This is actually a great way of seeing the topology of $\widehat{\mathbf{C}}$ (compare the end of §3.5): as we move out further in the complex plane, the preimages under S become more and more northerly. Thus, it is natural to say that $S(n) = \infty$, i.e., the north pole stereographically projects to ∞, and that $S : S^2 \to \widehat{\mathbf{C}}$ is a homeomorphism (tautologically giving the topology of $\widehat{\mathbf{C}}$, if you haven't already encountered it). So this map $\vec{N}(p) : \Sigma \to S^2$ is pretty clearly very geometric, and the map G is just the parametric and complex analytic rendering of it:

$$G = S \circ \vec{N} \circ \vec{u} : \mathcal{R} \xrightarrow{\vec{u}} \Sigma \xrightarrow{N} S^2 \xrightarrow{S} \widehat{\mathbf{C}}.$$

If you like vector calculus and computations, you should probably at this point check out whether I'm right or not: using (9) and inverse stereographic projection, find the unit normal map $\vec{N}$. (Finally, a small technicality that is buried here is that we must "orient" S^2 so that G preserves orientation.)

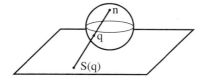

Figure 7. Stereographic projection.

3.8. Examples. After all this hard work to develop the Weierstrass representation (9), it's time to reap the rewards by just choosing functions G and dh.

Example 1. Enneper's surface (see Figure 8). The simplest choice (more about this in the next section §3.9) is to pick $\mathcal{R} = \mathbf{C}$, and $G = z$ and $dh = zdz$. This gives a complete immersed (but not embedded) surface known as Enneper's surface (1868). You can "build" it this way: take a coathanger bent to follow the seams on a baseball. Dip this in a soap solution to get a soap film. This film is *not* a portion of Enneper's surface. Break the film and dip it again; there is some chance that you'll get a different surface, but if you don't, then blow gently on the surface you do get and it will jump to a different surface which, in some sense, clings near to the other side of the baseball. The point is that there are two stable films that span this coathanger, so we expect an unstable surface on some path connecting them. (This is the notion of a "mountain pass": imagine on the landscape $\mathcal{L}$ of the space of all possible spanning surfaces, where height is given by area, that there are two valleys representing neighborhoods of the pair of stable surfaces. Then look at all the paths in this landscape of surfaces connecting the two valleys, and find the path whose maximum elevation is least. This is the path that takes you over the

Figure 8. Enneper's surface.
Figure by Matthias Weber.

Flat structures for minimal surfaces 103

mountain pass, and the point at its highest elevation represents an (unstable) minimal spanning surface to the wire — here it is a portion of Enneper's surface, if you properly bent the coathanger.)

Example 2. The catenoid; see Figure 2. Here take $\mathcal{R} = \mathbf{C} - \{0\}$, $G = z$ and $dh = \frac{dz}{z}$. Note the function-theoretic symmetry $z \mapsto \frac{1}{z}$ in the data that mirrors the reflection of the catenoid about the plane through its waist.

Example 3. The helicoid; see Figure 5. Here take $\mathcal{R} = \mathbf{C} - \{0\}$, $G = z$ and $dh = \frac{idz}{z}$. This one is a bit funny. Note that

$$\begin{aligned} Z &= \operatorname{Re} \int_{p=1}^{z} dh \\ &= \operatorname{Re} \int_{p=1}^{z} \frac{idz}{z} \\ &= \operatorname{Re}\{i \log z\} \quad \text{by our choice of } p = 1 \\ &= \operatorname{Re}\{i \log re^{i\theta}\} \\ &= \operatorname{Re}\{i \log r - \theta\} \\ &= -\theta. \end{aligned}$$

So as our path wraps around the origin, we steadily climb, and so this surface is not well-defined on our domain of $\mathbf{C} - \{0\}$. Yet it is still minimal: it simply has the topology (as the image surface $u(\mathcal{R}) \subset \mathbf{E}^3$ described in §2) of $\mathbf{R}^2$ and not of $\mathcal{R} = \mathbf{C} - \{0\}$, as our representation suggests. In our next section, we consider more carefully the perils of our defining forms α, β, and γ in (9) having integrals, around closed paths in $\mathcal{R}$, whose real parts do not vanish.

3.9. Some restrictions. After this beautiful formula (9) and the easy examples in §3.8, we come to the bad news: you just can't pick $\mathcal{R}$, G and dh arbitrarily. There are some restrictions — one quite severe, and one mild.

The mild restriction is the explanation of the use of the word "almost" in the paragraph after (7). The surface $\Sigma = u(\mathcal{R})$ has induced on it from $\mathbf{E}^3$ an element ds of arclength. (If this language is unfamiliar, think of the ways of measuring arclength from first year calculus, where a length of a curve γ in the plane was given as $\int_\gamma ds$. This is so useful and flexible that we extend its use to surfaces in space. By this we mean that if one defines an element of arclength ds on a surface (formally, a tensor that is a non-zero vector and returns a positive number, i.e., the absolute of a one-form), then one can define the length of an arc γ by integration, i.e., the length $\ell(\gamma) = \int_\gamma ds$. Since $\Sigma = u(\mathcal{R})$ is defined via the Weierstrass representation (9), we should be able to express this element of arclength in terms of G and dh. This expectation leads us to compute the expression

$$(10) \qquad ds = \frac{1}{\sqrt{2}} \left(|G| + \left| \frac{1}{G} \right| \right) |dh|$$

or, in terms of a coordinate z,

$$ds = \frac{1}{\sqrt{2}} \left(|G(z)| + \left| \frac{1}{G(z)} \right| \right) \left| \frac{dh}{dz} \right| |dz|.$$

We want this element ds of arclength to be well-behaved, i.e., neither 0 nor ∞. Thus, if dh has a zero, say of order n, at a point $p \in \mathcal{R}$, then we will need G to also have a zero or pole of order precisely n at that same point p; conversely, if G has a zero or pole of order n at $p \in \mathcal{R}$, then we will need dh to have a zero of that same order n at $p \in \mathcal{R}$.

This restriction, which we will refer to as the requirement that G and dh have *compatible divisors*, is mild, in that it only involves values of G and dh near isolated points, i.e., is a *local* condition on G and dh.

The real problem in constructing Weierstrass representations of minimal surfaces (where some aspect of the geometry or topology of Σ is controlled) is a *global* problem known as the *period problem* by the workers in the subject. As I describe it, keep the example of the helicoid in mind, because from a certain perspective, the helicoid has its periodic shape only because of this "period problem".

Flat structures for minimal surfaces 105

Put simply, the issue is that we want the map $u : \mathcal{R} \to \mathbf{E}^3$ to be well-defined, i.e., for each point $p \in \mathcal{R}$, there should be exactly one point $P \in \mathbf{E}^3$ for $p \in \mathcal{R}$ to map to via u. This is of course second nature to us: one of the first pieces of mathematical theory most of us learned is that functions are rules, and so are unambiguously defined. Yet look again at the definition of $u : \mathcal{R} \to \mathbf{E}^3$ in the Weierstrass representation:

(9)
$$u : z \longmapsto \left(\operatorname{Re} \int_p^z \frac{1}{2} \left(\frac{1}{G} - G \right) dh, \operatorname{Re} \int_p^z \frac{i}{2} \left(\frac{1}{G} + G \right) dh, \operatorname{Re} \int_p^z dh \right).$$

Here p is just some arbitrarily chosen point, and we can see that since along the trivial path from p to itself, we have $\int_p^p \alpha = \int_p^p \beta = \int_p^p \gamma = 0$, we must be defining $u(p) = (0,0,0)$. But what happens if the path Γ from p to itself is not just the trivial constant path? What if it loops as a cycle in some homologically non-trivial way? (Here a path Γ is "homologically non-trivial" if it is not the boundary of a subsurface — the idea behind invoking this hypothesis is to avoid situations where Γ bounds a subsurface $\Sigma^* \subset \Sigma$. This allows us to avoid situations where we might apply Stokes' theorem to the subsurface to find that $\int_\Gamma \delta = 0$ for any one-form δ which when restricted to Σ^* is then also holomorphic on that surface Σ^*.) If Γ were to be a non-trivial loop, then we could have problems, because, for example, it might happen that $\int_\Gamma \frac{1}{2} \left(\frac{1}{G} - G \right) dh$ is not purely imaginary, i.e., $\operatorname{Re} \int_\Gamma \frac{1}{2} \left(\frac{1}{G} - G \right) dh \neq 0$. In that case, we would have no unambiguous definition of $u(p)$; by parametrizing Γ as a path $\Gamma(t)$, say from $t = a$ to $t = b$, we would obtain a path $u(\Gamma)$ by writing

$$u(\Gamma(t)) = \left(\operatorname{Re} \int_a^t \frac{1}{2} \left(\frac{1}{G} - G \right) dh, \operatorname{Re} \int_a^t \frac{i}{2} \left(\frac{1}{G} + G \right) dh, \operatorname{Re} \int_a^t dh \right),$$

but that path would not close up in $\mathbf{E}^3$, since at least the first coordinate $\operatorname{Re} \int_a^b \frac{1}{2} \left(\frac{1}{G} - G \right) dh$ of the image at the end of the path would differ from zero, and the first coordinate of the image at the outset of the path would differ from the first coordinate of the image at the end of the path.

Thus the *major restriction* of a Weierstrass representation $u : \mathcal{R} \to \mathbf{E}^3$ is the requirement that

(11)
$$\operatorname{Re} \int_\sigma \frac{1}{2} \left(\frac{1}{G} - G \right) dh = 0,$$
$$\operatorname{Re} \int_\sigma \frac{i}{2} \left(\frac{1}{G} + G \right) dh = 0,$$
$$\operatorname{Re} \int_\sigma dh = 0$$

for any homologically non-trivial cycle $\sigma \subset \mathcal{R}$. See Figure 9.

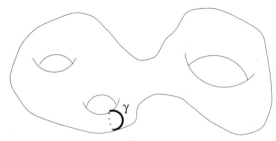

Figure 9. The global problem with the Weierstrass representation is the well-definedness around homologically non-trivial cycles.

Notice, for example, that the helicoid, represented as in Figure 5 in §2, violates the third item in (11) for any cycle that winds nontrivially around the origin in $\mathcal{R} = \mathbf{C} - \{0\}$.

4. Some recent non-trivial examples

4.1. The problem and some history. While the Weierstrass representation is more than a century old, the subject of minimal surfaces in space is quite young. Why? The reason is that the 19th century mathematicians didn't solve (either for lack of technique or lack of interest) the period problems on Riemann surfaces $\mathcal{R}$ with some handles. They also didn't know how tractable some of the function theory would turn out to be; see, e.g., [**Oss64**], [**Cos84**], [**HM82**]).

Flat structures for minimal surfaces

In the rest of the lecture, I would like to solve for you a particular problem in minimal surface theory that I recently worked on with a colleague, Matthias Weber, from Indiana University (then at Universität Bonn). I will of course only sketch the proof, but you can find full details (assuming a substantial background in differential geometry, minimal surfaces and complex analysis) in [**WW98**]. More precisely, I will fairly carefully describe the architecture of the proof, as the architecture involves many common elements of arguments in contemporary geometric analysis, but I'll get much vaguer at the end, when the details become a bit too particular to this case.

Before I begin, let me recall again some very brief historical context for our work. Remember that startling history: here is a list of the first few known complete embedded minimal surfaces which are not formed by repeating a basic pattern. The first is the plane, known since antiquity of course. The second is the catenoid, discovered in the eighteenth century. The third was found in the early 1980's by a brilliant Brazilian graduate student named Celso Costa [**Cos84**]; see also [**HM90**]. It is a torus with two catenoid ends and one planar end, and it is already pictured in Figure 6. This spectacular find fueled an already burgeoning interest in the subject, and people began to try to see what examples they could create.

(For instance, Hoffman and Meeks quickly saw how to add additional handles, as pictured in Figure 10.)

In rough outline, most of the construction techniques followed a pattern. First, dream up, either through imagination or computer graphics or a combination of both, a good picture of a complete minimal surface you think might exist. Then, and this is a critical step, assume that the surface has lots of symmetry; indeed, assume enough symmetry so that if you looked at the quotient of the surface by the group of isometries, you would be left with a very, very simple surface — commonly a surface topologically covered by a three-punctured sphere. Then study carefully the complex function theory of that simple surface — this is often quite difficult, as you need to know fairly precise information to solve the period problem — and use that study to prove that there is a solution to the period problem.

Figure 10. A Hoffman and Meeks generalization of Costa's surface. Figure by Matthias Weber.

Using this general outline, many fascinating examples were created and many very interesting questions resolved, laying the groundwork for a young theory.

The surfaces I want to speak about today are novel from the point of view of applying that above technique, in that they have very few symmetries. They are interesting from a theoretical point of view because they are provably the least complex minimal surfaces of genus g among all minimal surfaces of genus g.

Let me explain this last point. It has been known for about a century that closed oriented surfaces are classified topologically (i.e., up to homeomorphism) by their genus (or number of handles). So, to every complete minimal surface in space, we can associate a topological invariant, its genus. There is also a natural geometric invariant: we have discussed in section 3.7 the Gauss map $\overrightarrow{N} : \Sigma \subset \mathbf{E}^3 \to S^2$ from the minimal surface $\Sigma \subset \mathbf{E}^3$ to S^2. It turns out that each point $\theta \in S^2$ is obtained as often as each other point $\theta' \in S^2$ (if you are careful to count points $\theta \in S^2$ properly in the image of N when $dN = 0$).

Flat structures for minimal surfaces

Call this number of times that G maps Σ over S^2 the degree d of $\vec{N}$. Here's our question:

Question 1. If $\Sigma \subset \mathbf{E}^3$ is a complete, immersed minimal surface of genus g in $\mathbf{E}^3$, what is the least possible degree d for its Gauss map $\vec{N}$?

It is known [**JM83**] from some geometric topology that

$$d \geq g+1,$$

so our question becomes

Question 2. Are there any complete minimal surfaces Σ in $\mathbf{E}^3$ whose Gauss map has degree $d = g+1$ exactly?

This is the question that Weber and I answered, and that I will discuss here. It is very important that I be very clear about the history: independently of us, a Japanese mathematician named S. Sato also answered this question, and his initial manuscript [**Sa96**] appeared several months before ours did (with a very different proof). Thus he gets first credit for the general theorem. Other mathematicians had earlier produced examples in genus $g = 1, 2$, and 3, so they get credit for those.

Theorem A. Chen-Gackstatter [**CG82**] $g = 1, 2$; Do Espirito-Santo [**DoE94**] $g = 3$; [**Th94**] Experimental Evidence for $g \leq 35$; S. Sato [**Sa96**]; Weber-Wolf [**WW98**]. *There is such a surface for each genus $g \geq 0$; it has one infinite end which is asymptotically congruent to Enneper's surface.*

Here "asymptotically congruent" means that one could place a copy of Enneper's surface near the end of one of the surfaces from the theorem, and the two ends would have a distance between them that goes to zero as one travels out either end.

We give a computer image of a possible solution surface in Figure 11.

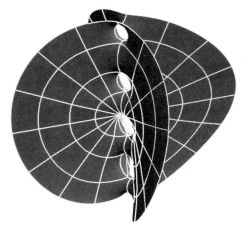

Figure 11. An image of a surface of genus 4 with one Enneper-like end. Figure by Matthias Weber.

4.2. The proof.

4.2.1. Revision of the period problem. In this section I want to begin the proof of Theorem A. First, we rearrange the algebra of the period conditions (11) to suggest a different approach to the Weierstrass representation (9). In particular, we observe that if we focus our attention on the pair of one-forms Gdh and $\frac{1}{G}dh$ (instead of the meromorphic function G and the one form dh), then we may express the period conditions (4) as

$$\int_\gamma Gdh = \overline{\int_\gamma \frac{1}{G}dh} \qquad \text{for all cycles } \gamma \subset \mathcal{R}$$

and
$$\operatorname{Re}\int_\gamma dh = 0 \qquad \text{for all cycles } \gamma \subset \mathcal{R}.$$

This seems counterproductive because now it seems that we need to track periods of Gdh, $\frac{1}{G}dh$, and dh as before, but in fact it will turn out that the third condition will be trivially satisfied, so this change in algebraic perspective really will clarify some issues for us.

Next comes a construction that will appear to be completely without motivation. In fact, I could explain how to pass from a picture of

Flat structures for minimal surfaces

a minimal surface you want to create to the following construction, but it would require me to develop a bit more complex analysis here than space allows. (You can find a pretty careful description in §2.3 of [**WW02**] that should be accessible to you if you know a semester of complex analysis theory.) It is possible, however, to discern much of the motivation from the discussion that follows the construction, especially if you review it backwards from where it ends, and I'll add a brief explanation at that time.

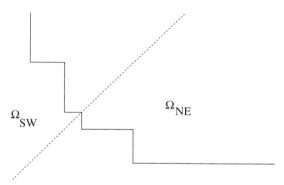

Figure 12. A staircase.

4.2.2. Staircases and flat surfaces with cone points. Here's the construction. Consider a polygonal arc in the plane with $2g + 1$ finite vertices that is symmetric with respect to reflection about the line $\{y = x\}$, and, as in Figure 12, alternates between left- and right-angled turns beginning from the positive y-axis. Call such a curve a "staircase" $\mathcal{S}$.

A staircase $\mathcal{S}$ divides the plane $\mathbf{C}$ into two complementary domains, which we will label as the northeast domain Ω_{NE} and the southwest domain Ω_{SW}:

$$\mathbf{C} \setminus \mathcal{S} = \Omega_{\mathrm{NE}} \cup \Omega_{\mathrm{SW}}.$$

Focus for example on the northeast domain Ω_{NE}, and think of it as a flat piece of $\mathbf{C}$, like a piece of steel, with a staircase-like boundary. In fact, take two copies of Ω_{NE}, the second oriented oppositely to the first, lay them exactly on top of each other and weld them together along the staircase boundary in $\widehat{\mathbf{C}}$ to get a new geometric object, S_{NE}.

What is S_{NE}? Well, if you interpreted the construction correctly and welded the two copies of the point ∞ to each other, then S_{NE} is topologically a sphere. But here the geometry is much more interesting than the topology: to begin, notice that the vertices of the stair case $\mathcal{S} = \partial \overline{\Omega}_{\mathrm{NE}}$ become, on the sphere S_{NE}, distinguished points, where somehow the geometry is different. *At all other points*, the geometry of a neighborhood of a point is the geometry of a neighborhood in a flat plane. This claim is pretty obvious at points in the interior of one of the old $\overline{\Omega}_{\mathrm{NE}}$'s, but it is also true for neighborhoods of non-vertices on $\partial \overline{\Omega}_{\mathrm{NE}} = \mathcal{S}$. You see, these points have neighborhoods on each $\overline{\Omega}_{\mathrm{NE}}$ that individually look like half a disk, and when you weld them together along their common boundary in $\partial \overline{\Omega}_{\mathrm{NE}}$, the union has the intrinsic geometry of a pair of welded half-disks, i.e., the intrinsic geometry of a flat disk. Thus, S_{NE} is a flat sphere with some distinguished points at the images of the vertices.

What about the vertices? Well, they are "cone points". To see this, let's examine a cone C of slant height 1. Cut C along a straight line L from its vertex to its boundary: the cone will now lay flat on a plane and will look like a disk with a sector of vertex angle θ removed, and the line L appears as the pair of boundary rays of that sector (see Figure 13). Thus the cone C is really the identification space of a sector of angle $\kappa = 2\pi - \theta$, where the identification is of the pair

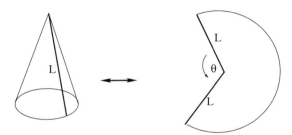

Figure 13. An extrinsic and synthetic view of a cone.

of boundary rays. Thus, there are two distinguishing features of this cone: it is an identification space of some "flat" Euclidean planar domains, and at the cone points, the circumference of a circle at distance ϵ from the singularity is $\kappa\epsilon$ where $\kappa \neq 2\pi$.

Back to the sphere S_{NE}: it is obviously the identification space of a pair of flat planar domains, and all of the finite vertices of the staircase $\mathcal{S}$ "lift" to S_{NE} as cone points with cone angles κ alternating between π and 3π. (A cone angle of 3π may seem odd, but you can manufacture a cone of angle 3π by simply identifying a pair of sectors of angle $3\pi/2$ and identifying pairs of boundary rays from opposite domains.) The cone angle on S_{NE} coming from the point at ∞ on Ω_{NE} is a different story. There the cone angle must be identified as $\kappa = -\pi$. This requires a bit of thought and a famous observation: the "total deficit" of cone angles from being 2π at every cone point on a sphere is 4π. To check this on examples, take a pair of congruent triangles in the plane and weld them together; if the original triangles had angles α, β and γ, then the welded sphere has cone angles $Z\alpha$, $Z\beta$ and $Z\gamma$, respectively. The total deficit is then $(2\pi - 2\alpha) + (2\pi - 2\beta) + (2\pi - 2\gamma) = 6\pi - 2(\alpha + \beta + \gamma) = 6\pi - 2\pi = 4\pi$. Then, to see the claim about the cone angle at ∞, take a quadrant and weld it to itself to find that cone angle at infinity for S_{NE} using the formula that the total angle defined should be 4π.

4.2.3. The surface cover and the lifted one-forms. There is one final step in the construction. Let M_{NE} denote the "two-fold branched cover of S_{NE}, branched over the cone points of S_{NE}." What is this? Well, there is a cut-and-paste way to understand this. The $2g + 2$ cone points sit along the equator of the sphere S_{NE}. Number them consecutively along the equator as $v_1, v_2, \ldots, v_{2g+2}$. Make a slit along the equator from v_1 to v_2, from v_3 to v_4, from v_5 to v_6, etc., so that the result, topologically, is a sphere with $g + 1$ holes in it. Take a second copy of this (thus the "two" in "two-fold branched cover") and glue the pair together by an identification that, for example, glues the upper lip of the slit from v_1 to v_2 on the first sphere to the lower lip of the slit from v_1 to v_2 on the second sphere, and also glues the other pair of lips of the slits from v_1 to v_2 together. Glue the other lips of the slits together similarly.

We conclude that, topologically, the surface M_{NE} is a closed orientable surface of genus g, i.e., a sphere with g handles. See Figure 14.

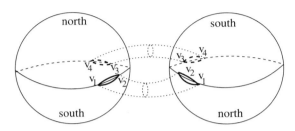

Figure 14. A surface of genus g.

The "branching" occurs at the vertices, because at those points, the construction yields just one image point, while at all other points P of S_{NE}, there are two copies of the point p on the surface M_{NE}; more to the point is that a circle $C^* \subset M_{\text{NE}}$ around the image point $v^* \in M_{\text{NE}}$ on M_{NE} of a cone point $v \in S_{\text{NE}}$ on S_{NE} has image $C \subset S_{\text{NE}}$ of a circle that maps twice around the vertex $v \in S_{\text{NE}}$. You should draw some abstract local pictures of the relationship of a neighborhood of a cone point on M_{NE} to the corresponding neighborhoods on S_{NE} to better understand that point.

Notice that M_{NE} inherits a flat cone metric from the pair of spheres S_{NE}, with the cone angles at the cone points doubling from what they were on S_{NE}.

Why do we do all this? Well, there are very few one-forms that people can claim to "know" completely; probably, we would say that we are very comfortable with the form dx on the real line $\mathbf{R}$ parametrized by $\{x \mid x \in \mathbf{R}\}$, or by extension, the one-form dz on the complex plane (sometimes called "the complex line") parametrized by the coordinate z. For instance, for the form dz, we know how to find all line integrals: if $\gamma_{AB} \subset \mathbf{C}$ connects A to B, then

$$(12) \qquad \int_{\gamma_{AB}} dz = B - A.$$

Let's briefly return to our discussion of one-forms, which we began in §3.6. I said then that one-forms eat vectors and return numbers:

Flat structures for minimal surfaces

this then means that one-forms are the natural tensor for integration along smooth curves, as at each point, they eat the tangent vector to the curve and the integral becomes the (generalized) sum of the resulting numbers. Now look at this one-form dz, restricted to the domain Ω_{NE} which is northeast of the staircase $\mathcal{S}$. This form "lifts" to the covering surface M_{NE} because M_{NE} is just made out of copies of Ω_{NE}, on which dz is defined (and after a bit of discussion, which I'll leave to you to fill in, about how the lift works near the images of $\mathcal{S}$). We'll call the resulting one-form, lifted from dz on Ω_{NE}, on M_{NE} by the suggestive name Gdh.

There's a similar construction for the "southwest side of $\mathcal{S}$", say Ω_{SW}; after copying it all out again, we find ourselves with a one-form, lifted from dz on Ω_{SW}, on the genus g surface M_{SW}. We'll call that one-form by the suggestive name $\frac{1}{G}dh$.

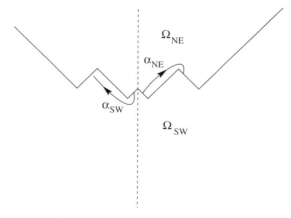

Figure 15. A rotated staircase and images of a cycle α.

After all of this constructing, here comes the payoff. It's easier to follow if we rotate $\mathcal{S}$ by $45°$ so that all of its sides now have slope either $+1$ or -1, and alternate between the two. See Figure 15. Pretend for a moment that M_{NE} is the same Riemann surface as M_{SW}, i.e., $M_{\mathrm{NE}} = M_{\mathrm{SW}} = \mathcal{R}$ (a Riemann surface), and let α be a cycle on $\mathcal{R}$. (We'll discuss at greater length what this equality really means in the next subsection. If we did it now, it would distract us from the basic plan.) Then α has images on $\mathbf{C}$ that might look as indicated in Figure 15. (As in Figure 15, assume that $\mathcal{S}$ has a reflective symmetry about a horizontal line through its middle vertex.) Then the two images $\alpha_{\rm NE}$ and $\alpha_{\rm SW}$ on $\Omega_{\rm NE}$ and $\Omega_{\rm SW}$ (respectively) are directed arcs that are conjugates of each other (after a rotation of $\Omega_{\rm SW}$ by π).

Then

$$\int_\alpha Gdh = \int_{\alpha_{\rm NE}} dz \qquad \text{as this is the definition of } Gdh \text{ and } \alpha_{\rm NE}$$

$$= \overline{\int_{\alpha_{\rm SW}} dz} \qquad \text{by construction of Figure 15 and (12)}$$

$$= \overline{\int_\alpha \frac{1}{G}dh} \qquad \text{by the definition of } \frac{1}{G}dh \text{ and } \alpha_{\rm SW}.$$

So we *seem* to have solved the period problem (11) for the minimal surfaces whose data G and dh combine to determine the Gdh and $\frac{1}{G}dh$ one-forms described here. In fact, those are the surfaces of Theorem A, where the Enneper-like end is represented by the vertex at infinity — to see this, one just studies pictures like Figure 11, and interprets the geometry of G and dh in terms of only a bit more complex function theory than we can describe here.

For the more advanced student, let me insert here a brief explanation, written with a slightly more advanced terminology. From pictures like Figure 11, we can determine all of the points with vertical normal vectors. If we identify via stereographic projection the upward pointing normal with ∞ and the downward pointing normal with 0, then we have found the divisor of G, at least at the finite points (i.e., not the ends) of the minimal surface. The Weierstrass data for the model of the ends, in this case the end of an Enneper surface for the asymptotically Enneper ends, determines the divisor at the points on the surface corresponding to the ends. (It is a wonderful and deep theorem of Osserman [**Os64**] that minimal surface ends of finite total curvature correspond to punctures on the underlying Riemann surface, and not holes or slits or anything else larger than a point.) Also, from the picture, or maybe by comparison with the divisor of G and using (10), we can determine the divisor of dh. From

Flat structures for minimal surfaces 117

these divisors, we can determine the divisors of the forms Gdh and $\frac{1}{G}dh$, using just the arithmetic rules for divisors. Now, the divisor of a meromorphic datum α, say $\alpha = Gdh$ or $\alpha = \frac{1}{G}dh$, determines that datum up to a constant; yet even with that ambiguity, we can still determine the flat singular metric $|\alpha|$, say $|\alpha| = |Gdh|$ or $|\alpha| = |\frac{1}{G}dh|$. That flat metric "develops" into the flat plane $\mathbf{E}^2$, by analytically continuing any initial local isometry of a patch into $\mathbf{E}^2$. The developed image Ω_α, like Ω_{Gdh}, carries a natural one-form dz, which pulls back by the developing map to the original one-form α. This is the procedure we secretly followed with one quarter of a surface such as that in Figure 11 to find the domains Ω_{Gdh} and $\Omega_{\frac{1}{G}dh}$. It is a bit of an accident that, in this case, the period conditions (10) force the two domains to fit together along their common boundary staircase $\mathcal{S}$.

4.2.4. The period problem gets rephrased as a mapping problem.

I said "we *seem* to have solved" in the last paragraph, rather than "we have solved," because there's still a big problem: in the discussion, we pretended that $M_{\mathrm{NE}} = M_{\mathrm{SW}} = \mathcal{R}$. But this is not necessarily so.

First observe that what we mean by "=" in $M_{\mathrm{NE}} = M_{\mathrm{SW}}$ is that (1) both M_{NE} and M_{SW} are Riemann surfaces, so that we have a way of measuring angles in both M_{NE} and M_{SW}, and (2) there is a homeomorphism from M_{NE} to M_{SW} which comes from taking Ω_{NE} to Ω_{SW} in a finite vertex-preserving way, and that homeomorphism is also angle-preserving.

The problem is, what if there is no such map? That is, what if M_{NE} and M_{SW} are topologically the same, but measure angles in a fundamentally different way?

In fact, this can happen. Probably by now you've seen the construction of a torus as an identified parallelogram, as in Figures 16. But in the square torus, the angle between the diagonal curves is 90° while in the rectangular torus, it is not 90°. One can prove, though it takes just a bit more than I'm providing here, that this angle between diagonals is a number that is characteristic of the Riemann surfaces represented here by the square torus and the rectangular torus (and their presentation here as a square and rectangle, but that's a bit more subtle, and we shall ignore this point completely). So the square torus and the rectangular torus are different Riemann surfaces.

We are left to consider the case of complementary domains Ω_{NE} and Ω_{SW} of a staircase $\mathcal{S}$. For most staircases $\mathcal{S}$, it turns out that the resulting Riemann surfaces M_{NE} and M_{SW} are different Riemann surfaces. More concretely, there is usually no conformal (i.e., angle-preserving) homeomorphism from Ω_{NE} to Ω_{SW} that takes finite vertices of $\mathcal{S}$ to finite vertices of $\mathcal{S}$.

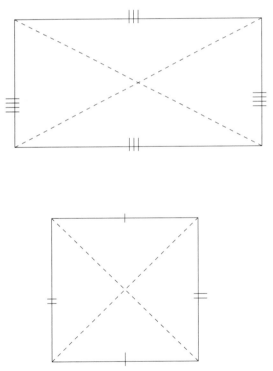

Figure 16. A square torus and a non-square torus.

Flat structures for minimal surfaces

But notice that we have lots of freedom in how we draw $\mathcal{S}$: thus Theorem A is proved by the above discussion if we can solve

Problem B. Find a staircase $\mathcal{S}$ (symmetric as in Figure 12) so that there is an angle-preserving map from Ω_{NE} to Ω_{SW} which takes finite vertices to finite vertices.

4.2.5. The moduli space of staircases.
Our new Problem B is phrased in a very common way if one adopts the point of view of what is known as moduli space theory. You see, each symmetric staircase with $2g+1$ finite vertices $\mathcal{S}$ can be ("almost") described in terms of $g-1$ real numbers, called moduli; we commonly refer to the space $\mathcal{M} = \{\mathcal{S}\}$ of symmetric staircases with $2g+1$ finite vertices as a "$(g-1)$-dimensional moduli space" of symmetric staircases.

There's a good chance that you've never heard of a "moduli space". That's fine — as I noted at the outset, much of my motivation for writing this article was that it afforded me the opportunity to introduce slightly advanced concepts such as moduli spaces well before they would naturally arise in standard coursework. Here's the idea: we are often interested in a collection of geometric spaces that are identical in some ways but not in all ways. Such a collection is called a moduli space, and invariants that we use to distinguish elements of the collection are called the moduli of those elements. The most famous moduli space is perhaps Riemann's moduli space of Riemann surfaces of genus g: this is the set of closed (boundaryless) Riemann surfaces all of which have the same genus (number of handles), but are different in that for any pair of distinct elements in the space — remember that this paper represents two Riemann surfaces of the same genus — there is no angle-preserving map between those surfaces. That is, each point on the space represents all the Riemann surfaces of genus g for which there is an angle-preserving map, but two different points really measure angles differently.

There are many other interesting (and famous) moduli spaces in geometry and mathematical physics. It is often interesting to study what shape these moduli spaces have and what geometries they admit, as that often reflects back to the geometry of the elements of the space. In fact, at the very end of this article, we will reduce our problem to the study of a certain function on a moduli space. We will end up being very interested in how the moduli spaces of small staircases fit into the boundary of moduli spaces of large staircases.

We return now to our basic problem. Before we try to solve Problem B, let's get the count right — explaining the word "almost" above — for there's a slightly subtle point here. Remember that our real interest is in the domains Ω_{Gdh} and $\Omega_{\frac{1}{G}dh}$, which are complementary to a symmetric staircase $\mathcal{S}$, and in fact, we only care about the way Ω_{Gdh} records angles — the way one measures distances in it is irrelevant. So if we start with a staircase $\mathcal{S}$ and then rescale it by a constant, say 7, to get a new staircase $\mathcal{S}' = 7\mathcal{S}$, then since the new complementary domains Ω'_{Gdh} and $\Omega'_{\frac{1}{G}dh}$ are just rescalings of the old Ω_{Gdh} and $\Omega_{\frac{1}{G}dh}$ (respectively), we really haven't changed the angle-measuring properties of the complementary domains. So we don't want to distinguish between $\mathcal{S}$ and $\mathcal{S}'$ in our moduli space. More trivially, we don't want to distinguish between congruent staircases, i.e., those differing by a rotation or translation.

It turns out that this scaling is the only way that two non-congruent staircases can give rise to equivalent complementary domains, where here "equivalent" means that there is an angle-preserving homeomorphism between Ω_{Gdh} and Ω'_{Gdh} and another angle-preserving homeomorphism between $\Omega_{\frac{1}{G}dh}$ and $\Omega'_{\frac{1}{G}dh}$. So our moduli space $\mathcal{M}$ should be thought of as a space of equivalence classes of staircases, where an equivalence class $[\mathcal{S}]$ consists of all those staircases $\mathcal{S}'$ which are a combination of rotations, translations and rescalings of $\mathcal{S}$.

So let's draw a staircase $\mathcal{S}$ and keep track of how much freedom we have. Up to rotation and translation, we can assume that the first finite vertex is 0, with the staircase containing the positive y-axis. The next finite vertex is on the real axis, so we can rescale it so that it occurs at the point 1. After that, we have normalized as much as possible, so the next $g - 1$ finite vertices are determined by any choice of $g - 1$ distances. Thus, we have determined the first $g + 1$ of the $2g + 1$ finite vertices, culminating in the choice of the "middle vertex". But all of the remaining vertices are determined by the symmetry about the 45° line through that middle vertex; thus the staircase is determined by $g - 1$ positive real numbers.

Flat structures for minimal surfaces

Thus $\mathcal{M}$ is a very simple moduli space, in that it is topologically trivial, and even naturally parametrized by $\mathbf{R}_+^{g-1} = \{(x_1, \ldots, x_{g-1}) \in \mathbf{R}^{g-1} \mid x_i > 0\}$.

So how does this perspective help us solve Problem B? Well, instead of trying to write down the Weierstrass data for Theorem A, we use very soft methods that we learned in calculus (these ideas are also preparatory to Robin Forman's lecture in this volume). We create a function $H : \mathcal{M} \to \mathbf{R}$ on $\mathcal{M}$ that somehow measures, for a given staircase, how close Ω_{Gdh} and $\Omega_{\frac{1}{G}dh}$ are to admitting the appropriate angle-preserving map; alternatively, this function measures how badly the angles in Ω_{Gdh} might be distorted by a good mapping from Ω_{Gdh} to $\Omega_{\frac{1}{G}dh}$.

We then prove three features of this "height" function $H : \mathcal{M} \to \mathbf{R}$.

1) We prove that $H(\mathcal{S}) \geq 0$ and $H(\mathcal{S}) = 0$ if and only if $\mathcal{S}$ solves Problem B.
2) We show that if $\mathcal{S}_n$ is a sequence of staircases that is not contained in a compact set $K \subset \mathcal{M}$, then $H(\mathcal{S}_n) \to \infty$.
3) We show that H is a differentiable function, and that if $H(\mathcal{S}) \neq 0$, then $\nabla H \neq 0$.

If we can engineer the creation of such a function $H : \mathcal{M} \to \mathbf{R}$, then Problem B is solved: Property (2) would guarantee the existence of a staircase $\mathcal{S}_0$ with $\nabla H(\mathcal{S}_0) = 0$, and then property (3) would imply that $H(\mathcal{S}_0) = 0$. By property (1), we would see that $\mathcal{S}_0$ solves Problem B.

At this point, I'd like to show you the construction of the height function $H : \mathcal{M} \to \mathbf{R}$, and the detailed proofs of (1)–(3), but this requires too substantial a background in complex analysis and moduli space theory. Besides, my purpose in this chapter was just to expose you to several contemporary fields and constructions in modern geometry, and the details of the proofs would take us in a different direction — into the techniques of a subject called Teichmüller theory. Of course, while you can read about the details of the proofs in [**WW98**], I bet it's a bit unsatisfying after such a long journey to get not even a glimpse of the whole picture. So I'll conclude this article by steering a middle course: I will give a vague summary of the construction and the techniques of proof, but not give a complete description. I hope that will be a bit more satisfactory.

Basically, there is a notion of a length of a class of curves that makes sense on a Riemann surface. This may be a bit confusing, since we said earlier that Riemann surfaces have only a well-defined notion of angle, not length, but the idea of the "extremal length" of a class of curves gets past that by considering a "best" length among all possible versions of lengths defined by metrics on the Riemann surfaces whose measurement of angles is compatible with that of the Riemann surface.

It turns out that if we pick the right $g-1$ curves, then these extremal lengths completely determine the domain Ω_{Gdh}, or the domain $\Omega_{\frac{1}{G}dh}$. The function H is defined to compare corresponding extremal lengths on Ω_{Gdh} and $\Omega_{\frac{1}{G}dh}$, and we can arrange the comparison so that $H \geq 0$. For most choices of these $g-1$ curves, the domains Ω_{Gdh} and $\Omega_{\frac{1}{G}dh}$ will solve problem B if and only if $H=0$, and so this proves property (1). If the curves intertwine properly, then property (2) will follow (surprisingly) from the fact that Ω_{Gdh} turns left at a vertex if and only if $\Omega_{\frac{1}{G}dh}$ turns right at the corresponding vertex. (The argument is long and technical, but that lack of correspondence is the crucial feature.) Finally, property (3) follows because of the way the domains fit together, and crucially from induction (!) on g. Basically, if we know we have a solution for genus $g-1$, then (by using the Implicit Function Theorem) we can add a small corner at the middle vertex of S and get a staircase for genus g. What's important about this genus g staircase is that it almost solves $H=0$; because it was born out of a solution for the $g-1$ problem, and $g=(g-1)+1$, all but one of its relevant extremal lengths are already in agreement. Thus, we find ourselves with but a one-parameter problem which is easily solved by moving a *simple* edge just back or forth. The tricky part in this final step is understanding how moduli spaces fit together. Certainly, since we can think of the genus g moduli space $\mathcal{M} = \mathbf{R}_+^{g-1}$, we can think of the genus $g-1$ moduli space as a face $\mathbf{R}_+^{g-2}$ of $\mathbf{R}_+^{g-1} = \mathcal{M}$. But it is a bit more technical to understand how the important functions on $\mathbf{R}_+^{g-2}$ extend to $\mathbf{R}_+^{g-1}$. I think I have to leave those details to the paper, though.

Flat structures for minimal surfaces 123

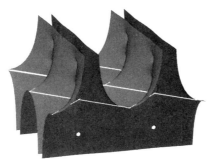

Figure 17. Doubly-periodic Scherk surface
whose quotient surface has genus 4.
Figure by Matthias Weber.

Figure 18. A Costa-esque surface with a top catenoid end, two
middle planar ends, a bottom catenoid end, and genus 4.
Figure by Matthias Weber.

Let me conclude with some comments about where this work led
Weber and me. It turned out that in the years between the time

this conference was held and this paper made it to final form, we understood our proof much better. We also saw how, with some modifications, we were able to use it to add handles to surfaces other than Enneper's surface. In Figures 17 and 18, there are images of the doubly periodic Scherk surface with handles, and the Costa surface with additional handles and flat ends. A present challenge, now that we are beginning to understand how to prove the existence of some of these surfaces, is to understand what properties characterize them uniquely.

Finally, a substantial breakthrough occurred in 2001, when Martin Traizet [**Tr02**] produced an example of a complete embedded minimal surface with no symmetries at all.

References

[Cos84] C. Costa, *Example of a complete minimal immersion in $\mathbb{R}^3$ of genus one and three embedded ends*, Bull. Soc. Bras. Mat. **15** (1984), 47–54.

[CG82] C.C. Chen and F. Gackstatter, *Elliptische und hyperelliptische Funktionen und vollständige Minimalflächen von Enneperschen Typ*, Math. Ann. **259** (1982), 359–369.

[DHKW92] U. Dierkes, S. Hildebrandt, A. Küster, and O. Wohlrab, *Minimal Surfaces I, II*, Springer-Verlag, 1992.

[DoE94] N. Do Espírito-Santo, *Superfícies mínimas completas em $\mathbb{R}^3$ com fim de tipo Enneper*, PH.D. Thesis (1992) (published as: *Complete minimal surfaces in $\mathbb{R}^3$ with type Enneper end*, Ann. Inst. Fourier (Grenoble) **44** (1994), pp. 525–577), University of Niteroi, Brazil.

[HT85] S. Hildebrandt and A. Tromba, *Mathematics and Optimal Form*, Scientific American Library, 1985.

[HM90] D. Hoffman and W.H. Meeks III, *Embedded Minimal Surfaces of Finite Topology*, Ann. of Math. **131** (1990), 1–34.

[JM83] L. Jorge and W.H. Meeks III, *The topology of complete minimal surfaces of finite total Gaussian curvature*, Topology **22(2)** (1983), 203–221.

[Mo85] F. Morgan, *An Introduction to Geometric Measure Theory*, Academic Press, 1985.

[Oss64]	R. Osserman, *Global properties of minimal surfaces in E^3 and E^n*, Annals of Math. **80(2)** (1964), 340–364.
[Oss86]	R. Osserman, *A Survey of Minimal Surfaces*, Dover Publications, 1986.
[Sa96]	K. Sato, *Existence proof of one-ended minimal surfaces with finite total curvature*, Tohoku Math. J. (2) **48** (1996), 229–246.
[Str88]	M. Struwe, *Plateau's Problem and the Calculus of Variations*, Princeton University Press, 1988.
[Tr02]	M. Traizet, *An embedded minimal surface with no symmetries*, J. Differential Geom. **60** (2002), 103–153.
[Th94]	E. Thayer, *Complete Minimal Surfaces in Euclidean 3-Space*, Univ. of Mass. Thesis, 1994.
[WW98]	M. Weber and M. Wolf, *Minimal Surfaces of Least Total Curvature and Moduli of Plane Polygonal Arcs*, Geom. and Funct. Anal., **8** (1998), 1129-1170.
[WW02]	_____, *Teichmüller Theory and Handle Addition for Minimal Surfaces*, Annals of Math. (2) **156** (2002), 713-795.

Hold That Light!
Modeling of Traffic Flow by Differential Equations

Barbara Lee Keyfitz

1. Introduction: A continuum model for traffic flow

I begin this paper with a disclaimer: I am not a traffic engineer, nor an expert in the use of the model I am about to present to you. However, the basic model here, developed by Lighthill and Whitham [22], and by Richards [27], is well-regarded, at least by academic traffic engineering researchers. It has been presented in a number of elementary and advanced textbooks, for example [6, 9, 21, 29], as well as in introductions to mathematical modeling [5, 11]. The expository article by Gazis [8] briefly mentions this model, and the model is still used in traffic engineering research to treat certain situations such as bottlenecks, as in a technical paper by Newell [24].

My motivation in this paper is to show you something of the way applied mathematics works, by deriving a differential equation from physical principles and common sense, solving the equation, and then interpreting the answer as it refers to the phenomenon being modeled. In the example here, the equation can be solved by elementary

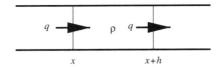

Figure 1. The control section for one-way traffic

methods: we do not get a closed-form solution, but we get enough information to visualize the solution.

There is a second reason for introducing the traffic flow model. It illustrates a widely-used approach to dynamic problems, an approach which often leads to systems of partial differential equations in the form of conservation laws. The approach is to act as though a quantity which one wants to monitor is continuously distributed in space, rather than being a discrete variable. In the case of my example here, the quantity is cars on a highway — traffic — which is quite obviously discrete. However, for the sake of argument, suppose that we are not interested in the motion of individual cars, but only in some averaged quantities — for example, the carrying capacity of the road, which is the maximum number of cars per hour which the road can accommodate, or the number of times a car keeping up with the traffic will be stopped by a red light on a given stretch of road. In cases like this, one gets a good approximation, which appears to agree with the data, by considering, instead of individual cars, a continuously distributed quantity, ρ, the *linear density* of traffic on the road, measured in cars per mile. We suppose there is only one road, that it is straight and uniform, and that traffic moves along it in one direction (left to right): a straight one-way highway with no intersections. Other embellishments can be added.

Now, the equation describing the dynamics of traffic on this road is very simple to derive: it is the *law of conservation of cars*, which is a statement that the way the amount of traffic on a section of the road changes in time is simply by a net flux, that is, the difference between the rate at which traffic enters the section of road and the rate at which traffic leaves. We may take the road to be the x-axis, and examine a control section between x and $x + h$, as in Figure 1.

Hold That Light!

Assume the density is a function of x and of time, denoted by t. Since the amount of traffic in the control section is the integral of the density, the fundamental equation is

$$(1) \qquad \frac{d}{dt} \int_x^{x+h} \rho(y,t)\, dy = q(x,t) - q(x+h,t).$$

The quantity q is the *linear flux* of traffic. Flux may not be as familiar a concept as density. In this example, it is natural to think of it as the number of cars passing the point x in a unit of time. Of course, we are not looking at individual cars, so one should properly think of the units of q as the amount of traffic per hour, measured in cars per hour.

We have a simple way of relating the flux to more familiar quantities in traffic: if the speed of the traffic is v, in units of miles per hour, then the amount of traffic passing a given point x is just ρv cars per hour. Of course, if you think of actual traffic, composed of a lot of vehicles of different sizes, all traveling at different rates of speed, then v is a composite quantity which might be difficult to calculate. But in this simple model, just as there is a single quantity which represents density, so there is a single velocity, $v(x,t)$, at each point x and time t.

Applying the mean value theorem for integrals to the left side of (1), we obtain

$$\frac{\partial}{\partial t}\big(h\rho(x^*,t)\big) = q(x+h,t) - q(x,t),$$

where the change in notation from an ordinary to a partial derivative with respect to t follows convention. The value x^* is a point in the interval $(x, x+h)$. Divide by h to get

$$\frac{\partial}{\partial t}\big(\rho(x^*,t)\big) = \frac{q(x+h,t) - q(x,t)}{h},$$

and finally take the limit $h \to 0$ to obtain the fundamental conservation principle in the form of a partial differential equation:

$$(2) \qquad \frac{\partial \rho}{\partial t} + \frac{\partial q}{\partial x} = \frac{\partial \rho}{\partial t} + \frac{\partial (\rho v)}{\partial x} = 0,$$

where we have substituted $q = \rho v$ to get the second form.

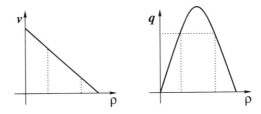

Figure 2. Idealized velocity and flux curves

Now, the reasoning here was straightforward, and, as long as you accept that ρ represents an averaged quantity, and v the velocity that corresponds to it, we have not made any approximations: the total amount of traffic is exactly conserved, and that is all equation (2) says.

On the other hand, we have obtained only an incomplete model, because (2) is a single equation in two unknowns, ρ and v. For partial differential equations, as for any other kind of equation, you cannot find a satisfactory solution unless the number of equations is the same as the number of unknowns. It is an interesting fact that modeling with conservation laws almost always brings one to this same place: it is straightforward in many physical situations to reason that certain quantities are conserved — mass, momentum, energy, and so on — but it always turns out that the number of unknowns is greater than the number of equations. In this example, the only physical quantity which is conserved is mass, which is why we obtain a single equation. A comparison with fluid dynamics modeling is given in Section 4.

To obtain a solvable model from (2), one invokes a *closure* assumption. To be specific, in this case we assume that the velocity is a function of the density, that is, $v = v(\rho)$. This is a statement that, on this particular road, the speed at which traffic moves is completely determined by its density. Such a model is an oversimplification of actual traffic, as it does not take into account the variations of individual drivers or different kinds of vehicles. However, when we examine what the model predicts, we find it quite realistic in some respects.

Hold That Light!

There is no theoretical basis to determine the shape of the curve $v(\rho)$. The following postulate is intuitively reasonable: v is a monotone decreasing function of ρ, with a maximum at $\rho = 0$. At some value ρ_0, v becomes zero: ρ_0 is a critical density above which traffic cannot move. For single lane traffic, this has been measured at about 225 cars per mile [**29**, page 68]. The flux, $q(\rho) = \rho v(\rho)$ is thus zero at $\rho = 0$ and $\rho = \rho_0$. Some experimental data are quoted by Haberman [**11**, page 286]. A number of functional relationships have been proposed to approximate this empirical relationship; see [**9**, page 57]. We shall assume, because it makes the problem much simpler and because it is what the data show, that q is concave down: $q'' = 2v' + \rho v'' < 0$. We get a qualitative picture (this model is due to Greenshields [**9**]) by letting v decrease linearly with increasing ρ; we can choose units of measurement so that $v = 1 - \rho$ and $q = \rho - \rho^2$, as in Figure 2. The illustrations and examples will use this function. The theory, however, is exactly the same for any $v(\rho)$ which leads to a concave flux q.

Notice a feature of any concave flux function: to each value of q below the maximum, there correspond two values of density, one lower and one higher, and two values of velocity, one higher and one lower. That is, a road can process a given volume of traffic in one of two modes. At the lower density and greater vehicle speed, an individual driver will complete the trip more rapidly, and this is considered preferable by most of us. Section 3 gives an example of how a traffic configuration can slip from the more desirable mode into the other.

Now, with $q(\rho)$ a known, differentiable function of ρ, equation (2) becomes an example of a *first-order quasilinear partial differential equation in conservation form*, otherwise known as a *scalar conservation law*. Equations of this form were examined in the 1940s and 1950s for several reasons. First, Jan Burgers [**4**] had proposed a related equation as a model for the formation of coherent structures in turbulent fluids. Burgers' equation has the form

$$(3) \qquad \frac{\partial u}{\partial t} + \frac{\partial}{\partial x}\left(\frac{1}{2}u^2\right) = \nu \frac{\partial^2 u}{\partial x^2},$$

which differs from (2) in that the flux function is convex rather than concave — this is an inessential difference, as one can replace u by $-u$ — and, more substantially, in having a second-order term on the right-hand side. The parameter ν in Burgers' model is proportional to the viscosity of the fluid. Burgers was interested in features of turbulence which appeared at very small viscosities, and so considered the case $\nu = 0$, which is just our traffic flow equation after a simple change of variables. A mathematical theory for scalar conservation laws was developed by Hopf [**12**], Lax [**19, 20**], Oleĭnik [**25**] and others.

In the next section, we will use this example to develop properties of solutions of first-order partial differential equations and of conservation laws. The mathematical properties have appealing interpretations as features of traffic flow.

2. Some conservation law theory

As a first step, we rewrite (2), carrying out the differentiation:

$$(4) \qquad \frac{\partial \rho}{\partial t} + \frac{\partial q(\rho)}{\partial x} = \frac{\partial \rho}{\partial t} + a(\rho)\frac{\partial \rho}{\partial x} = 0.$$

Now $a(\rho) = dq/d\rho$ is another known function of the density, called the *wave speed*. Let us see why.

Consider first an unrealistic special case: suppose our traffic has the property that v is independent of ρ: that is, $v = v_0$ for all ρ and $q = \rho v_0$. Then $a = v_0$ and the equation becomes

$$(5) \qquad \frac{\partial \rho}{\partial t} + v_0 \frac{\partial \rho}{\partial x} = 0.$$

This is a *linear* equation (which means simply that the principle of superposition holds and the solutions form a vector space). The following proposition is easily demonstrated.

Proposition 2.1. *Every differentiable solution of* (5) *is of the form*

$$\rho(x,t) = \rho_0(x - v_0 t)$$

for a differentiable function ρ_0 of one variable. Furthermore, every choice of ρ_0 yields a solution of the equation.

Hold That Light!

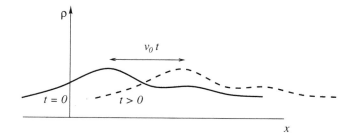

Figure 3. Density profiles for the linear equation

Proof. After a change of independent variables
$$y = x - v_0 t, \qquad s = t,$$
the equation becomes
$$\frac{\partial \rho}{\partial s} = 0,$$
so the solution in the new coordinates must be a function of y alone. Direct differentiation shows that every function of y satisfies the equation. □

What information about traffic is conveyed by this solution? The message is that any pattern of densities simply moves to the right with speed v_0, as in Figure 3. The function ρ_0 corresponds to a distribution of densities along the road at a particular time; for example, if $\rho_0(x)$ is the density at position x at time $t = 0$, then $\rho(x,t) = \rho_0(x - v_0 t)$ is the solution of the *initial value problem* consisting of equation (5) and the initial condition

(6) $$\rho(x, 0) = \rho_0(x), \qquad -\infty < x < \infty.$$

Another way of picturing the way traffic patterns change is by means of a space-time diagram, as in Figure 4. The lines $x - v_0 t = $ const. are contours of constant density. What Proposition 2.1 shows is that the lines $x - v_0 t = $ const. play a special role in solving the equation. These distinguished curves are called *characteristics* for the equation.

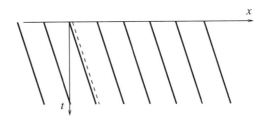

Figure 4. Characteristics for the linear equation

We return now to equation (4). This equation is not linear; rather, it is *quasilinear*, a word which refers to the fact that it is linear in the derivative terms, $\partial \rho/\partial t$ and $\partial \rho/\partial x$, but the coefficients may depend on ρ. The *method of characteristics* can be used to solve this equation, by seeking curves $x(t)$ with special properties. In fact, if we differentiate ρ with respect to t, assuming x to be a function of t, we obtain

$$\frac{d}{dt}\rho\bigl(x(t),t\bigr) = \rho_x \frac{dx}{dt} + \rho_t$$

(where subscripts denote partial derivatives) and now, comparing with (4), we see that if we choose $x(t)$ to satisfy $dx/dt = a(\rho)$, then this becomes

$$\frac{d}{dt}\rho\bigl(x(t),t\bigr) = \rho_t + a(\rho)\rho_x = 0\,,$$

so that ρ is constant along the curve $x(t)$, which then must be a straight line (since its reciprocal slope is $a(\rho)$, which is also constant). In fact, if (x_0, t_0) is any point on the line, and the value of the density is ρ_0 there, then the equation of the line is $x(t) = x_0 + a(\rho_0)(t - t_0)$, and $\rho = \rho_0$ at every point on the line. If we suppose a function $\rho_0(x)$ given by means of an initial condition (6), then $t_0 = 0$ and we can write the solution by means of an equation

(7) $$\rho\bigl(x_0 + a(\rho_0(x_0))t, t\bigr) = \rho_0(x_0),$$

which gives $\rho(x,t)$ implicitly as long as we can solve the equation

(8) $$x = x_0 + a\bigl(\rho_0(x_0)\bigr)t$$

to find x_0 as a function of x (for fixed t). This we can do if x is a monotonic increasing function of x_0 (again for a fixed value of t).

Hold That Light!

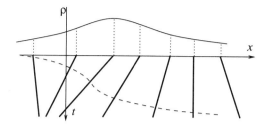

Figure 5. Characteristics for the quasilinear equation

Now,
$$\frac{dx}{dx_0} = 1 + a'\rho_0' t,$$
by the chain rule, where $a' = a'(\rho_0(x_0))$ and $\rho_0' = \rho_0'(x_0)$. At $t=0$ the right side of the equation equals unity, and so the expression remains positive, at least for small t.

Although $\rho(x,t)$ cannot be found explicitly as in the linear case, it is possible to visualize how the densities evolve with time by means of a diagram analogous to Figure 4. Given an initial density distribution, $\rho_0(x_0)$, from each point x_0 on the x-axis draw the straight line (8), as in Figure 5. Notice that $a(\rho_0)$ is the slope of the line drawn in a conventional (t,x)-plane (that is, with the t-axis horizontal). The convention in partial differential equations is to make the x-axis the horizontal axis; a further convention in traffic engineering is to draw the t-axis downward (so you can see a as the slope if you rotate the picture). For reference, the initial density distribution that was used to draw the picture is sketched above the x-axis: it contains a density wave, with a peak value near the maximum value $\rho = 1$, in the center. The concentration is initially symmetric front to back, decreasing to a value $\rho = 1/3$ at the edges of the picture. If we assume $q(\rho) = \rho - \rho^2$, so $a(\rho) = 1 - 2\rho$, then the slopes dx/dt vary between $1/3$ and -1. The evolution of the density wave with time is shown in Figure 6. One sees that the wave remains a wave, but it no longer preserves its shape. The peak moves more slowly than the wave as a whole; in fact, the peak may even move backwards. After some time, a wave that was originally symmetric, front to back, has become skewed. It is steeper

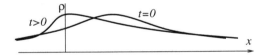

Figure 6. Density profiles for the quasilinear equation

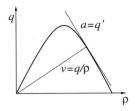

Figure 7. Vehicle velocity and wave velocity

at the back (engineers call this a compression wave) and slopes more gently at the front (this is called a rarefaction). Remembering the assumption that denser traffic moves more slowly, we see that the picture is a simple consequence of the modeling assumption. A patch of traffic in which congestion is greater at the rear will gradually stretch out, as the cars at the front race away, while a situation in which low-density traffic is spread out behind a slow-moving traffic jam will tend to become more concentrated as the cars in the back catch up with the crowd.

It is time to say a word about what is moving in the wave. In our first, linear model, the line $x = x_0 + v_0 t$ in space-time represented both the position at any time of the car which had been at x_0 at the initial time, and the position of the point on the wave (peak, front, back) as it evolves in time. Such is no longer the case in the quasilinear model. In fact, every point in the wave with density $\rho > 1/2$ is actually moving backwards in time, but every car is moving forward or at worst standing still! Recall that at a given density ρ, the velocity of a car is $v = q/\rho$, in terms of the flux function q, while the velocity of that part of the wave (that is, the part with value ρ) is $a = q' = v + \rho v'$, which is less than v if q is not constant. See Figure 7, in which both

Hold That Light!

speeds are depicted as slopes. This brings up an interesting point, which is central to the study of wave motion: what travels when a wave moves? Think, for example, of the waves that fans make in a sports arena: the wave travels around the arena, but the arms that generate it move up and down. It might be reasonable to say that what is traveling is information. Indeed, traffic modeling that takes into account individual drivers and their reaction times reaches the same conclusion about wave motion by postulating that drivers react to a patch of dense traffic just ahead by slowing down (thus contributing to increasing density of traffic and passing the signal on to drivers further in the rear). It is not difficult to trace the route of an individual driver through the density wave shown in Figures 4 and 5. A sample space-time path is shown by a dotted line. Wave motion in which points in different parts (or phases) or the wave have different velocities is said to have a *dispersive* character (Whitham [**29**]).

However, examination of the situation in Figure 5 for times beyond the end of the drawing shows that the theory — in particular, the implicitly given solution (7) — will fail eventually. For the straight lines in Figure 5 are not parallel, and so they must intersect. (The sole exception is the case of data where all the lines are spreading apart.) Fortunately, this does not mean a collision, since the lines are not the trajectories of individual cars, but it does mean that the mathematical function obtained by the method of characteristics no longer solves the problem. The resolution of this difficulty is interesting, and is at the heart of the subject of conservation laws. To understand it, one must go back to the beginning of the derivation of the model, to the calculus assumption under which the differential equation (2) was derived from the integral equation (1). In writing the differential equation we assumed, without stating this explicitly, that ρ and q were differentiable functions of x and t. But there are *discontinuous* functions for which (1) holds. For example, the piecewise constant function

(9)
$$\rho(x,t) = \begin{cases} \rho_0, & x < st \\ \rho_1, & x > st \end{cases}, \quad q(x,t) = \begin{cases} q_0, & x < st \\ q_1, & x > st \end{cases}, \quad s = \frac{q_1 - q_0}{\rho_1 - \rho_0}$$

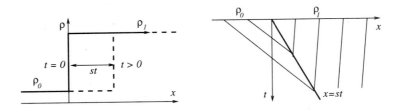

Figure 8. Piecewise constant solution; profile and characteristics

solves equation (1) exactly. This function also represents a traffic configuration; its profile is shown in Figure 8, along with the space-time diagram of the characteristics in the case $\rho_0 < \rho_1$ (we shall see shortly why this is the only interesting case).

What is the interpretation of a discontinuous solution like this? The discontinuity, called a shock, is a demarcation line between regions of heavy and of light traffic. For example, if cars are backed up behind a red light, then $\rho_0 = 1$ and $\rho_1 = 0$; the corresponding fluxes in our model are $q_0 = q_1 = 0$, so $s = 0$ and traffic sitting at a light is a solution to the equation. It is conventional to refer to discontinuous solutions as *weak* solutions, because they do not satisfy the differential equation (2) in the classical sense, but instead satisfy the equation (1). (An even more general notion of a weak solution is conventionally used, but will not be needed in this paper.) The situation $0 < \rho_0 < \rho_1 \le 1$ also admits an all-too-familiar interpretation: cars free-wheeling on the expressway suddenly come upon a patch of dense slow (or stopped) traffic, slam on their brakes and join the queue. Depending on the relative densities of cars before and behind the shock, the discontinuity may move forward or backward, or stand still.

As the graph of the characteristics in Figure 5 makes clear, one does not expect there to exist a differentiable solution of (4) for all $t > 0$ if the initial data look like a wave with a density maximum in the center. In fact, the implicitly defined solution given by equations (7) and (8) ceases to exist at the first time where $dx/dx_0 = 0$, which is the smallest positive value of $-1/(a'\rho_0')$. Typically, the profile ρ develops a vertical tangent at this value of t, which grows into a shock

Hold That Light! 139

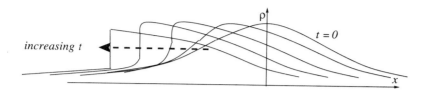

Figure 9. Development and decay of a shock

for later t (the precise development of this singularity is interesting and was studied by a number of people; for a recent description, see Nakane [23]). Far away from the shock, the method of characteristics still gives a valid solution, but after a while more and more characteristics run into the shock and terminate; the size of the jump in ρ also grows initially, but then stops growing and begins to decrease, as sketched in Figure 9. (The asymptotic form of the wave for large t, often called an N-wave, is given by Lax, [20].)

We now consider a final feature of solutions of conservation law equations, which must be resolved by mathematical theory. Let us return to the initial-value problem with piecewise constant data

$$\rho(x,0) = \begin{cases} \rho_0, & x < 0 \\ \rho_1, & x > 0 \end{cases}$$

with $\rho_0 > \rho_1$. This type of data is called *Riemann data* in the field of conservation laws. (We saw that $\rho_0 = 1$ and $\rho_1 = 0$ might represent cars stopped at a red light. One could set $t = 0$ at the moment the light turns green. You can imagine an interpretation for other values of ρ_i: perhaps a flagman has been slowing the traffic and has just walked away.) We saw that the piecewise constant function ρ given in equation (9) is a solution. But when $\rho_0 > \rho_1$ there is a second solution, as you can check:

$$(10) \quad \rho(x,t) = \begin{cases} \rho_0, & x < (1 - 2\rho_0)t, \\ \frac{1}{2}(1 - \frac{x}{t}), & (1 - 2\rho_0)t < x < (1 - 2\rho_1)t, \\ \rho_1, & x > (1 - 2\rho_1)t. \end{cases}$$

The solution profile and characteristic lines are shown in Figure 10. This fanlike solution is called a *centered rarefaction wave*. It is an idealized portrait of an everyday phenomenon: when a line of congested

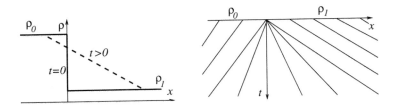

Figure 10. A centered rarefaction wave

traffic is faced with an emptier road ahead, the congestion breaks up as the first cars race off into the open space; the opportunity to accelerate gradually moves back through the line.

Mathematically, the existence of two solutions for the same initial condition is a sign of *ill-posedness*: the problem is badly or incompletely specified. In fact, much modeling, especially of complicated situations, results in ill-posed problems, and much effort in applied mathematics goes into finding satisfactory resolutions of the dilemmas which result. An *ad hoc* fix is available here: reasoning from the situation faced when a light turns green, one sees that the two solutions represent different cases: the traffic remains at the intersection (the shock wave), or begins to move (the rarefaction). The second is what actually happens (the 'physical' solution). A good deal of research has gone into finding what are called *admissibility conditions* for shocks — conditions which will, ideally, be strong enough to eliminate spurious solutions but not so restrictive as to eliminate all solutions.

Peter Lax has given a condition which tells which discontinuities are admissible for our model (the condition can also be extended to systems of conservation laws). Called the *geometric entropy condition*, Lax's condition states that shocks are admissible if the characteristics on either side run into the shock in the direction of forward time. Thus, the shock in Figure 8 is admissible, but a shock drawn in the situation of Figure 10 would not be, because the characteristics on either side flow away from each other. A heuristic argument for why this is a reasonable condition is that, under this condition, every point (x, t) in space-time, except points that are actually on shocks, can be connected to the initial data by a unique characteristic line, and hence the solution is uniquely given there. As long as most points are not on shocks, then this is enough to determine a weak solution, since the integral equation satisfied by a weak solution is not affected by the values of ρ at a discrete set of points.

To justify the model, one wants a theory that gives both existence and uniqueness. For the traffic-flow model, and in fact for any scalar conservation law, such a theory exists. It was first developed by Lax [19, 20] for a convex flux function and extended by Oleĭnik [25]. When q is neither convex nor concave, new phenomena appear; see also [13]. A complete theory for a single equation was given by Kružkov [17] in 1970. There is also extensive theory for systems; the first existence theorem was given by Glimm [10] in 1965. Recent work by Bressan and co-workers [3] is finally tackling questions of uniqueness, but many problems remain open.

Let me state, without proof, Lax's theorem for the system examined in this paper.

Theorem 2.1. *The solution to*

$$\rho_t + (\rho - \rho^2)_x = 0, \qquad \rho(x,0) = \rho_0(x),$$

for $t > 0$ is given by

$$\rho(x,t) = \frac{1}{2}\left(1 - \frac{x - y_0}{t}\right)$$

where y_0 is the value of y at which the function

$$G(x,y,t) = \int_{-\infty}^{y} \rho_0(s)\, ds + \frac{1}{2}\left(x - y - \frac{(x-y)^2}{2t}\right)$$

achieves its minimum. Provided that the integral of ρ_0 exists, this formula gives the solution at almost every point (x,t).

It is straightforward to verify that this formula gives the function (7) defined by the method of characteristics, for times before the first shock forms. One can also check that any shock solution given by this formula will satisfy Lax's condition. Lax's proof in [20] that the formula gives a solution for all time, and for all integrable data, uses a connection between the traffic-flow equation and Burgers' equation, (3).

At this point one can establish, via several steps, a connection with the calculus of variations. First note that if we define a new quantity u by $u_x = \rho$ or

$$u = \int_{-\infty}^{x} \rho(s,t)\, ds,$$

then, whenever ρ satisfies equation (4), u will satisfy

(11) $$u_t + q(u_x) = 0.$$

An equation of this form is called a *Hamilton-Jacobi* equation. The calculus of variations provides an elegant method for solving (11), from which the formulas of Theorem 2.1 can be derived. An excellent exposition can be found in the text of Evans [7]. The last third of Evans' book also contains extensions of this example problem. The relation with Burgers' equation, (3), plays a role in the theory of viscosity solutions of Hamilton-Jacobi equations, as is also discussed in [7].

3. An application of the model: The timing of traffic lights

One of things this model can do is predict, again under rather idealized conditions, what will happen at a red light, and how lights should be timed, depending on how heavy the traffic is. It also explains why, for an individual driver, there is no perfect staging of lights. The calculation was done by Richards [27], one of the originators of the model, and has been reproduced since, for example in Haberman [11] and Whitham [29].

Let us consider, first, a single traffic light which is red for a time interval T_R and then green for an interval of duration T_G. The density of traffic everywhere on the road can be calculated explicitly if we make a few simplifying assumptions: assume that the incoming density has the constant value ρ_0 and that the velocity-density relation is linear, $v = 1 - \rho$. Assuming that traffic is running free when the light changes, the red cycle is described by the first picture in Figure 11: cars back up behind the light, and a shock forms and moves up

Hold That Light! 143

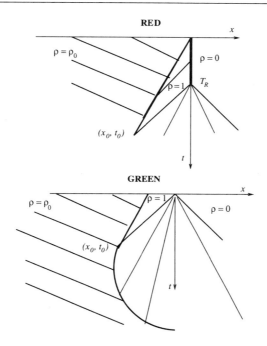

Figure 11. The traffic light cycle

the street with speed

$$s = \frac{q_1 - q_0}{\rho_1 - \rho_0} = 1 - (\rho_0 + \rho_1) = -\rho_0.$$

When the light turns green, traffic begins to move at the light, and a rarefaction forms; if we take this instant to be the origin of time, then the rarefaction is centered at the origin, and has head and tail speeds 1 and -1 respectively. The tail of the rarefaction runs into the shock at

$$t_0 = \frac{\rho_0}{1 - \rho_0} T_R, \qquad x_0 = -\frac{\rho_0}{1 - \rho_0} T_R,$$

and now the speed of the shock changes. If we let $\phi(t)$ denote the position of the shock, then the shock speed is $s = d\phi/dt$ and ϕ satisfies the differential equation

$$\frac{d\phi}{dt} = 1 - \left(\rho_0 + \frac{1}{2}\left(1 - \frac{\phi}{t}\right)\right)$$

since ρ_1, the density to the right of the shock, is given by equation (10). Rearranging terms gives a linear, first-order ordinary differential equation

$$\frac{d\phi}{dt} - \frac{\phi}{2t} = \frac{1}{2} - \rho_0,$$

which we can solve. Applying the initial condition $\phi(t_0) = x_0 = -t_0$ we get

$$\phi(t) = (1 - 2\rho_0)t - 2\sqrt{\rho_0(1-\rho_0)T_R t}$$

as the equation of the arc the shock describes in space-time. If $\rho_0 \geq \frac{1}{2}$, that is, if the road is at greater than half its carrying capacity, then the shock continues to move to the left for all time. Thus, even if the light never turns red again, the effect of the delay propagates back up the road forever. (The influence does, however, decay.)

On the other hand, if $\rho_0 < \frac{1}{2}$, the shock speed eventually becomes positive, and the shock crosses the point $x = 0$ at a time

$$t_r = \frac{\rho_0(1-\rho_0)T_R}{(\frac{1}{2} - \rho_0)^2},$$

provided the light is still green.

It turns out to make a big difference whether the shock gets through the intersection or not before the light changes again. The discovery of this effect, and its quantification (which can be improved from the calculation here by taking a more realistic velocity-density equation) are a major contribution of the model. From a mathematical viewpoint, this example illustrates a difference between nonlinear and linear models. Linear problems "scale up": if the effect of doubling the density is to multiply the waiting time at a light by a factor K, then quadrupling the density multiplies it by $2K$, and proportionally smaller changes will have smaller effects. But in this model, the effect of increases in the density is small until a threshold is reached, after which the nature of the traffic flow changes dramatically.

To see how, and why, look again at the density-flux diagram, Figure 2, and recall that a desired traffic flux which is less than the

Hold That Light!

maximum value of q can be realized at two different densities. From the point of view of the road, the service factor is the same, but individual drivers will notice a great difference in the time it takes them to make the trip.

Let us first suppose that the incoming density, ρ_0, and the timing of red and green lights are such that the shock crosses the intersection before the light turns red again; that is, $T_G \geq t_r$. If we assume that the cycle of lights is fixed, this means that

$$\rho_0 \leq \frac{1}{2}\left(1 - \sqrt{\frac{T_R}{T_R + T_G}}\right). \tag{12}$$

Since the density behind the shock is again ρ_0 when the light turns red again, then in this situation after the red-green cycle we have the same initial condition as before, and to the left of the light the flow is periodic in time. It is also periodic beyond the light, and the average density a short distance downstream must also be ρ_0: the effect of the light is to impose on the traffic a temporary slowdown, from which it recovers.

However, let us now suppose that the shock does not get through the intersection. We can make the following calculation of the flux at the light itself: during the green cycle, the density is precisely $\frac{1}{2}$ (since it is the center of the rarefaction) and so is the velocity, while during the red cycle the flux is zero, so the average flux over a period is

$$\bar{q} = \frac{1}{4}\frac{T_R}{T_R + T_G}.$$

Now, this must be the average flux anywhere along the road (since there are no sources or sinks), and so the average density is found by solving $\rho - \rho^2 = \bar{q}$. This equation has two roots, but the smaller is precluded by the fact that inequality (12) is violated. Hence we have

$$\bar{\rho} = \frac{1}{2}\left(1 + \sqrt{\frac{T_R}{T_R + T_G}}\right)$$

as the value of the average density. The effect of a small change in incoming density from a value satisfying (12) to one just violating it has been to move the average density from the lower to the upper

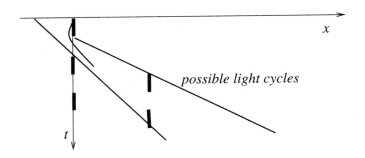

Figure 12. A road with more than one light

half of the density-flux diagram, greatly increasing the travel time for drivers.

The actual numbers are dramatic: Richards [**27**] points out that if $T_R = T_G$, then the greatest incoming density that permits free flow at the light is $(1 - 1/\sqrt{2})/2 \approx .15$, while the average density when the shock does not get through the light is $(1 + 1/\sqrt{2})/2 \approx .85$.

A second feature of roads with traffic lights, also familiar to drivers, can be derived from this model. This has to do with the timing of successive lights along a road. If we look at the flow behind a well-timed light, then it is periodic in time, with period $T_R + T_G$. However, beyond the light the mass that has crossed the light in a single green cycle (which corresponds to a set of drivers, as well) spreads out, since the velocities vary between 1 and $1 - 2\rho_0$. See Figure 12. If there is a second light on the road, and its cycle has the same period, then the only way to arrange its cycle so that the entire mass gets through the second light on a single cycle is to make the ratio T_G/T_R greater for the second light. Even then, this works only if the second light is sufficiently close to the first. Realistically, a series of lights will be timed so that only a part of this packet will get through later lights. Which part? Reasoning on the basis of the fluxes, it is optimal to plan so that the later part of the packet, where the density is greater, will cross the second light without waiting. Thus it is to be expected that the first drivers through a green light will get caught at the next light. A light that is timed for the benefit of the first phalanx of

Hold That Light! 147

drivers will have the effect of cutting off the last drivers through the previous light, as in the title of this paper.

4. Extensions and other models

One source of interest in the continuum traffic flow equation is its resemblance to the *compressible Euler equations* used to study the dynamics of gases. A simple version of the equations, for one-dimensional flow, is

$$\rho_t + (\rho v)_x = 0,$$
$$(\rho v)_t + (\rho v^2 + p)_x = 0.$$

The first equation represents conservation of mass for a gas with density ρ and velocity v, and it is exactly the same as the traffic flow equation, except that we no longer assume v to be a function of ρ. The second equation is conservation of momentum and is the expression, in a continuum, of Newton's law of dynamics, $F = ma$. (Models in which v is postulated to be a function of ρ, instead of being determined from a second equation, are called *kinematic*, as contrasted with *dynamic*.)

One sees, in this system, that there is again a closure problem: the two equations contain a third variable, the pressure, p. Under some conditions (roughly that there is not much energy exchange in the flow), one can establish p as a function of ρ from thermodynamic principles. One then has a system which can be written in the form

(13) $$\mathbf{u}_t + \mathbf{f}(\mathbf{u})_x = 0$$

where $\mathbf{u}$ is the vector of states, $\mathbf{u} = (\rho, \rho v)$, and $\mathbf{f}$ the flux vector. This system can also be written in quasilinear form,

(14) $$\mathbf{u}_t + A(\mathbf{u})\mathbf{u}_x = 0,$$

where A is the Jacobian matrix, $\partial \mathbf{f}/\partial \mathbf{u}$. Just as for the single equation, it is useful to look at the linear problem one obtains on replacing A by a constant or by a known function of x and t.

When do the methods established in the previous sections help in solving equation (13) or (14)? The system behaves like the scalar equation in important ways precisely when it is *hyperbolic*, that is, when the matrix A has real eigenvalues and is diagonalizable. In fact, it is a simple exercise in linear algebra to show that if A is constant and $P^{-1}AP = D$ is diagonal for a constant matrix P, then in a new coordinate system, $\mathbf{w} = P^{-1}\mathbf{u}$, equation (14) separates into a set of uncoupled equations, each of the form of the scalar linear equation, (5), with an eigenvalue of A playing the role of velocity. The eigenvalues are called *characteristics* for the system, and using this change of coordinates one can write down the general solution for this linear, constant-coefficient system. In the nonlinear case, the system does not decouple in any coordinate system, and there is no analogue to the implicit formula (7) for the solution to an initial-value problem. However, one can show that in many ways the solutions to a system behave like solutions to the equation we have studied: waves steepen, shocks form, and disturbances propagate through the flow, with velocities equal to one of the characteristic speeds.

It may not be immediately apparent that a system with complex eigenvalues behaves differently. For example, Laplace's equation, $\Delta v \equiv v_{xx} + v_{yy} = 0$, when written as a first-order system for $\mathbf{u} = (v_x, v_y)$ is a coupled linear system with eigenvalues $\pm i$. Laplace's equation is a perfectly reasonable equation and serves as a model for many physical processes. However, none of these involves time explicitly; a difficulty arises, not with the equation itself but with the problem of trying to solve it along with data given on one axis, say $y = 0$. This prototype example is known as *Hadamard's counterexample* and can be found in most partial differential equations textbooks, for example, Evans [**7**, page 233]. It is simple to demonstrate that data of small amplitude and high frequency (functions like $\sin nx/n$) lead to solutions of very large amplitude for non-zero values of y (functions like $\sin nx \sinh ny$). In the limit $n \to \infty$ the data tend to zero but the solution does not. Thus, the solution does not depend continuously on the data, which is another way an equation can be ill-posed.

Extension from a scalar equation to a system of conservation laws, so important for gas dynamics, also occurs in modeling traffic flow. First, there have been attempts to find two-equation models for the one-way traffic flow problem we have been describing. This is

Hold That Light!

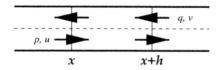

Figure 13. Variables in the two-way traffic model

motivated partly by a desire to find solutions that agree better with actual traffic. In addition, the velocity-density relation which closed the single equation model does not have a theoretical justification, and people have sought a more scientific way of deriving it. A recent paper of Aw and Rascle [1] surveys the progress of this attempt and offers a new suggestion. A related use of two-equation models is to provide a reduction to the single-equation model by the use of what are called relaxation methods, as is done by Lattanzio and Marcati [18]. (A different attempt find $v(\rho)$ was made by Prigogine and Herman [26], using ideas from statistical mechanics and kinetic theory. Their book is an engaging introduction both to traffic theory and to statistical mechanics.)

However, I would like to close this paper by mentioning a very simple two-equation model, which is closely related to the one-way model we have been discussing. Bick and Newell [2], almost forty years ago, proposed to extend the continuum model to two-way traffic on a road, by supposing that a density for traffic in each direction can be assigned, with a corresponding flux function, and that the two directions of traffic interfere with each other in some way.

Here is the two-way traffic model. (See Figure 13.) Changing notation somewhat, we let $p(x,t)$ and $q(x,t)$ be the densities of east-bound and west-bound traffic, respectively, and let u and v be the respective velocities. We assume that there are no U-turns, so conservation of mass holds for each direction of travel separately, and we obtain the system

$$\begin{aligned} p_t + (pu)_x &= 0, \\ q_t + (qv)_x &= 0. \end{aligned}$$

If we assume that each lane completely ignores the other and acts as a one-way road, then the closure assumptions from before are appropriate:

$$u = u(p) = 1 - p, \qquad v = v(q) = -(1 - q),$$

where v is negative since the west-bound traffic moves in the direction of decreasing x. This system is hyperbolic; in fact, the characteristic speeds are $1 - 2p$ and $2q - 1$. The system is not *strictly hyperbolic*; that is to say, the speeds are not everywhere distinct. They coincide when $p + q = 1$; if we think of (p, q) as a point in a *phase space*, then the characteristics coincide on a line in two-dimensional phase space. However, the system remains hyperbolic there, since there is no eigenvector deficiency.

Now, the idea explored by Bick and Newell was to suppose that the two lanes of traffic interfere with each other. No specific mechanism need be mentioned; perhaps passing is inhibited, perhaps a sense of crowding discourages higher speeds. To study the effect of this assumption, we can look at the simplest type of dependence:

$$u = u(p, q) = 1 - p - \beta q, \qquad v = v(p, q) = -(1 - q - \beta p),$$

where β is a small coupling parameter. The equations are coupled if $\beta > 0$. In addition, the feasible region of phase space, which had been a square $\{0 \leq p \leq 1, 0 \leq q \leq 1\}$ now becomes a quadrilateral bounded by $p = 0$, $q = 0$, $p + \beta q = 1$, and $q + \beta p = 1$. But the most significant difference is that the coupled system is not hyperbolic for all states: computing the eigenvalues of the Jacobian matrix $A(p, q)$ shows that there is a region near the line $p + q = 1$, elliptical in shape, with width of the order of β, in which the eigenvalues of A are complex conjugates. Bick and Newell performed some studies on this system; they identified some of the shocks and solved some Riemann problems. However, they could not explain away this surprising feature, nor solve the problem for all initial conditions. More recently, Vinod, while a graduate student at the University of Houston, explored this problem further, and showed that in a number of cases one could find solutions for $\beta > 0$ which approached the easily calculated solutions for the uncoupled case, $\beta = 0$, [**28**].

It is surprising that the two-way model, which seems like a simple and natural extension of the one-way model, turns out to have unexpected mathematical features. It is, at one level, completely ill-posed, in the sense that the corresponding linear problem may suffer from the Hadamard instability. Indeed, a complete solution to the mathematical problem is not known at this time. I came across the model because it is related to research I have been doing on other systems of conservation laws, used in physics and engineering, which are, similarly, derived from what appear to be reasonable principles and from analogies to well-known and well-behaved models, and which are ill-posed in exactly the same way. This work is reviewed in [**14, 15, 16**].

A note of caution is in order here. Often one is presented with equations which are said to be mathematical models for something, and one is given the impression that there is not much depth to the modeling exercise — understand the physics well enough, and understand the mathematics well enough, and everything will be straightforward. In examples such as the two-way traffic model, the interplay between "modeling" and "mathematics" is intricate and subtle. The process goes both ways: as more is understood about the mathematics, people return to the models to take another look. The one-way traffic model was not useful to engineers until the mathematical theory was developed in the 1950s. Perhaps future theories will make sense of two-way traffic models.

Bibliography

[1] A. Aw and M. Rascle, Resurrection of 'second order' models of traffic flow. *SIAM J. Appl. Math.* 60 (2000), no. 3, 916–938.

[2] J. H. Bick and G. F. Newell, A continuum model for two-directional traffic flow. *Quarterly of Applied Mathematics*, XVIII:191-204, 1960.

[3] A. Bressan, *Hyperbolic systems of Conservation Laws: The One-Dimensional Cauchy Problem.* Oxford University Press, Oxford, 2000.

[4] J. M. Burgers, A mathematical model illustrating the theory of turbulence. *Advances in Applied Mathematics*, 1:171-199, 1948.

[5] D. A. Drew, Traffic flow theory. in M. Braun, C. S. Coleman, and D. A. Drew, (eds.), *Differential Equation Models*, volume 1 of *Modules in Applied Mathematics*, Chapter 14. Springer-Verlag, New York, 1983.

[6] D. A. Drew, *Traffic Flow Theory and Control*. McGraw-Hill, New York, 1968.

[7] L. C. Evans, *Partial Differential Equations*. American Mathematical Society, Providence, 1998.

[8] D. C. Gazis, Traffic flow and control: Theory and applications. *American Scientist*, 60:414-424, 1972.

[9] D. L. Gerlough and M. J. Huber, *Traffic Flow Theory*. Transportation Research Board, National Research Council, Washington, 1975.

[10] J. Glimm, Solutions in the large for nonlinear hyperbolic systems of equations. *Communications on Pure and Applied Mathematics*, 18:95-105, 1965.

[11] R. Haberman, *Mathematical Models: Mechanical Vibrations, Population Dynamics, and Traffic Flow*. Prentice-Hall, Englewood Cliffs, 1977.

[12] E. Hopf, The partial differential equation $u_t + uu_x = \mu u_{xx}$. *Comunications on Pure and Applied Mathematics*, 3:201-230, 1950.

[13] B. L. Keyfitz, Solutions with shocks: an example of an L^1 contractive semigroup. *Communications on Pure and Applied Mathematics*, XXIV:125-132, 1971.

[14] B. L. Keyfitz, Conservation laws that change type and porous medium flow: a review. In W. E. Fitzgibbon and M. F. Wheeler, (eds.), *Modeling and Analysis of Diffusive and Advective Processes in Geosciences*, pages 122-145. Society for Industrial and Applied Mathematics, Philadelphia, 1992.

[15] B. L. Keyfitz, Multiphase saturation equations, change of type and inaccessible regions. In J. Douglas, C. J. van Duijn, and U. Hornung, (eds.), *Proceedings of Oberwolfach Conference on Porous Media*, pages 103-116. Birkhäuser, Basel, 1993.

[16] B. L. Keyfitz, A geometric theory of conservation laws which change type. *Zeitschrift für Angewandte Mathematik und Mechanik*, 75:571-581, 1995.

[17] S. N. Kružkov, First-order quasilinear equations in several independent variables. *Mathematics of the USSR - Sbornik*, 10:217-243, 1970.

[18] C. Lattanzio and P. Marcati, The zero relaxation limit for the hydrodynamic Whitham traffic flow model. *Journal of Differential Equations*, 141:150-178, 1997.

[19] P. D. Lax, Weak solutions of nonlinear hyperbolic equations and their numerical computation. *Communications on Pure and Applied Mathematics*, VII:159-193, 1954.

[20] P. D. Lax, Hyperbolic systems of conservation laws, II. *Communications on Pure and Applied Mathematics*, X:537-566, 1957.

[21] W. Leutzbach, *Introduction to the Theory of Traffic Flow*. Springer-Verlag, Berlin, 1988.

[22] M. J. Lighthill and G. B. Whitham, On kinematic waves. II. A theory of traffic flow on long crowded roads. *Proceedings of the Royal Society*, A229:317-345, 1955.

[23] S. Nakane, Formation of shocks for a single conservation law. *SIAM Journal on Mathematical Analysis*, 19:1391-1408, 1988.

[24] G. F. Newell, Traffic flow for the morning commute. *Transportation Science*, 22:47-58, 1988.

[25] O. A. Oleĭnik, Discontinuous solutions of nonlinear differential equations. *American Mathematical Society Translations, Series 2*, 26:95-172, 1957.

[26] I. Prigogine and R. Herman, *Kinetic Theory of Vehicular Traffic*. American Elsevier, New York, 1971.

[27] P. I. Richards, Shock waves on the highway. *Operations Research*, 4:42-51, 1956.

[28] V. Vinod, *Structural Stability of Riemann Solutions for a Multiphase Kinematic Conservation Law Model that Changes Type*. Ph.D. thesis, University of Houston, Houston, Texas 77204, 1992.

[29] G. B. Whitham, *Linear and Nonlinear Waves*. Wiley-Interscience, New York, 1974.

Titles in This Series

26 **Robert Hardt,** Editor, Six themes on variation, 2004
25 **S. V. Duzhin and B. D. Chebotarevsky,** Transformation groups for beginners, 2004
24 **Bruce M. Landman and Aaron Robertson,** Ramsey theory on the integers, 2004
23 **S. K. Lando,** Lectures on generating functions, 2003
22 **Andreas Arvanitoyeorgos,** An introduction to Lie groups and the geometry of homogeneous spaces, 2003
21 **W. J. Kaczor and M. T. Nowak,** Problems in mathematical analysis III: Integration, 2003
20 **Klaus Hulek,** Elementary algebraic geometry, 2003
19 **A. Shen and N. K. Vereshchagin,** Computable functions, 2003
18 **V. V. Yaschenko,** Editor, Cryptography: An introduction, 2002
17 **A. Shen and N. K. Vereshchagin,** Basic set theory, 2002
16 **Wolfgang Kühnel,** Differential geometry: curves - surfaces - manifolds, 2002
15 **Gerd Fischer,** Plane algebraic curves, 2001
14 **V. A. Vassiliev,** Introduction to topology, 2001
13 **Frederick J. Almgren, Jr.,** Plateau's problem: An invitation to varifold geometry, 2001
12 **W. J. Kaczor and M. T. Nowak,** Problems in mathematical analysis II: Continuity and differentiation, 2001
11 **Michael Mesterton-Gibbons,** An introduction to game-theoretic modelling, 2000
10 **John Oprea,** The mathematics of soap films: Explorations with Maple®, 2000
9 **David E. Blair,** Inversion theory and conformal mapping, 2000
8 **Edward B. Burger,** Exploring the number jungle: A journey into diophantine analysis, 2000
7 **Judy L. Walker,** Codes and curves, 2000
6 **Gérald Tenenbaum and Michel Mendès France,** The prime numbers and their distribution, 2000
5 **Alexander Mehlmann,** The game's afoot! Game theory in myth and paradox, 2000
4 **W. J. Kaczor and M. T. Nowak,** Problems in mathematical analysis I: Real numbers, sequences and series, 2000
3 **Roger Knobel,** An introduction to the mathematical theory of waves, 2000
2 **Gregory F. Lawler and Lester N. Coyle,** Lectures on contemporary probability, 1999
1 **Charles Radin,** Miles of tiles, 1999